A guide to basic print production
Book 1 Planning the project

Judith Wilkinson

Practical ACTION PUBLISHING

The British Council in association with Intermediate Technology Publications

Acknowledgements

I would like to thank all those who have helped with planning and production friends and family, British Council colleagues, professional associates workshop participants and in particular the following:

Denise Ayres and her colleagues (Appropriate Health Resources and Technologies Action Group)
Bernard Battley (Battley Brothers Ltd, Printers)
Anne Devillebichot (Speaker Support Services)
Fay Edinborough (Grey Coat School, London)
Dick Fletcher (New Media Publishing)
Peter Hilken (The British Council, Kenya and Nigeria)
Henry Larken (London College of Printing)
Ian McLaren (Lanchester Polytechnic, Coventry)
Peter Mayer (Goldsmiths College, University of London)
Cliff Morris (University of Reading)
Shirley Parfitt (Consultant, Communications for Development)
David Warr (Consultant, Allama Iqbal Open University, Pakistan)
for their knowledge, advice, patience and goodwill. J.W.

Published by the British Council in association with
Intermediate Technology Publications Ltd
(trading as Practical Action Publishing Ltd)
25 Albert Street, Rugby, CV21 2SD, Warwickshire, UK
www.practicalactionpublishing.org

© The British Council, 1985

First published 1985\Digitised 2008

ISBN 13 Paperback: 9780946688661
ISBN 10: 0946688664
ISBN Library Ebook: 9781780442983
Book DOI: http://dx.doi.org/10.3362/9781780442983

All rights reserved. No part of this publication may be reprinted or reproduced or utilized in any form or by any electronic, mechanical, or other means, now known or hereafter invented, including photocopying and recording, or in any information storage or retrieval system, without the written permission of the publishers.

A catalogue record for this book is available from the British Library.

The authors, contributors and/or editors have asserted their rights under the Copyright Designs and Patents Act 1988 to be identified as authors of their respective contributions.

Since 1974, Practical Action Publishing has published and disseminated books and information in support of international development work throughout the world. Practical Action Publishing is a trading name of Practical Action Publishing Ltd (Company Reg. No. 1159018), the wholly owned publishing company of Practical Action. Practical Action Publishing trades only in support of its parent charity objectives and any profits are covenanted back to Practical Action (Charity Reg. No. 247257, Group VAT Registration No. 880 9924 76).

Illustrations by Sue Henry
Design by The British Council

Contents

5 **About This Book**

7 **Section A: Identifying the Nature of the Task**

9 **Chapter 1 Communication**
9 Communication as a Process
10 Source
10 Message
11 Channel
11 Audience
12 Effects
12 Feedback
13 Communication as Common Understanding
14 Problems in Communication

17 **Chapter 2 Audience and Message**
17 Finding Out about the Audience
18 How to Find Out
19 Steps in a Simple Survey
28 Preparing the Message
30 Deciding on the Topic
30 Studying the Topic

32 **Chapter 3 Aims and Objectives**
32 Aims
32 Objectives
33 Motivation

35 **Section B: Making Choices — Methods and Media**

37 **Chapter 4 The Range of Possibilities**
37 Factors in Making a Choice
37 A Multi-Media Approach
40 Methods and Media — An Overall View

41 **Chapter 5 Narrowing the Field**
41 Communication Methods
41 Learning through Achievement
42 Simulation: Models, Games, Drama, Puppetry
44 Story-telling
44 Song
45 Discussion
45 Field Trip
45 Demonstration

46	Talk or Lecture
46	Exhibition

47 Low-Cost Media

48	Sound Recordings
48	Visual Aids: Projected, Three-Dimensional, Printed Material

52	Case Study Exercise

57 Section C: Using Printed Material

59 Chapter 6 Why Use Printed Material?

59	What is Printing?
59	Advantages of Printed Material as a Medium
59	Disadvantages and Problems
60	Low-Cost Printing
60	Centralized and Local Production
61	Some Possible Alternatives

62 Chapter 7 Forming the Message

62	Contents
62	Information
64	Approach

65 Writing Simply

67 Choosing Illustrations

67	Picture Recognition
69	Picture Style
71	Symbols
73	Diagrams
75	Using Illustrations

79 Pretesting

83	Using a Questionnaire
86	Analysing Findings

90 Some Final Comments

91 Glossary

94 References

95 Bibliography

About This Book

This is the first in a series of four related books. It deals with the thinking, planning and preparation that should take place before production of material begins. Whatever kind of material it is, it will take time and money. Producers will therefore want it to be worth the effort, and to do the job it was meant to do. Careful planning is one way to help to achieve this.

Before actually starting work, you will need to find out a number of things. The three key questions are:

> - WHO is this material intended for?
> - WHAT do you intend to tell them?
> - WHAT results or reactions are desired as a result of your message?

These questions must be answered fully and accurately in order to establish the exact purpose of the proposed material. Once this is clear, it will be much easier to make efficient decisions about all the other points that have to be considered. This book looks at the three key questions, with a number of others, and explores ways of collecting information and analysing it in order to make your material as effective as possible.

Section A: Identifying the Nature of the Task

Section A looks at the audience, the message and its intended results and discusses some relevant aspects of communication. It explains why accurate background information is so important to communication, and how to find out what you need to know.

Section B: Making Choices — Methods and Media

Having established why material is needed and who it is for, ask: 'What exactly should I produce?' In order to decide what material will really be most effective, producers need to know about the alternatives available. This part of the book looks at the range of possibilities, and gives information about those methods and media that are likely to be available, at a realistic price, and therefore within the scope of most people using this package. This general outline is followed by a more detailed look at choosing the methods and media most appropriate for a given set of circumstances. It goes into the reasons why printed material is a popular choice of medium, and the characteristics that make it a valuable way to communicate a message.

Section C: Using Printed Material

Section C deals with printed material and its potential. It introduces choices you will have to make and things you will need to know about; it gives advice about writing simply, presenting information clearly, choosing appropriate illustrations and pretesting material. It thus leads into the different aspects of print production covered by the other books in the series.

Note

In communication theory, some everyday words are used with a precise meaning in mind. At the end of this book there is a glossary designed to help you with these words, and also with words that may be new to you.

Section A

Identifying the Nature of the Task

Chapter 1   Communication

Why do you need to use printed material at all? Here are some possible answers — do they give good reasons for producing material?

• 'One part of the campaign to preserve fertile soil is to warn farmers about the dangers of soil erosion; pictures can help with the task of informing and motivating.'

• 'Extension workers are finding it difficult to explain methods of contraception when they talk to villagers in their homes; some support material with simple diagrams could be useful to them.'

• 'That talk we give about immunization is never very well attended; it will make the information more interesting if we include some illustrated material.'

• 'We have a grant for media production — let's make some posters!'

• 'I've told the villagers that they should take more care about washing their hands, but they still don't seem to understand that dirt can spread disease; I need to reinforce this message; maybe I could make illustrated materials for them to look at while I'm away — I can discuss the subject again next time I call at the village.'

Notice that the answers marked *yes* are based on some problem in communication — a problem that people feel printed material or other media can help to solve. Communication in these examples means someone conveying an idea or message to an audience, for a particular reason — usually to bring about changes in their:

• knowledge of the idea

• feelings about the idea

• behaviour

or even all three. Communication and change are therefore very closely interwoven.

The 'someone' who wants to convey the message might be close to his audience, for example a health extension worker talking to a family in their home, or quite remote from where the action is taking place, for example a doctor whose written booklet is being used to train villagers in basic health care. Whoever he is, and wherever he is, he hopes to communicate effectively. In order to do this, he must be able to answer the three key questions mentioned on page 5 and to define the intended audience, the message to be conveyed and the desired results or reactions.

Communication as a Process

It would help at this point to look more closely at what 'communication' might be. There are different ways of simplifying and explaining this complex subject, but it is often seen as the process by which an idea is transferred from a source to an audience or 'receiver'.

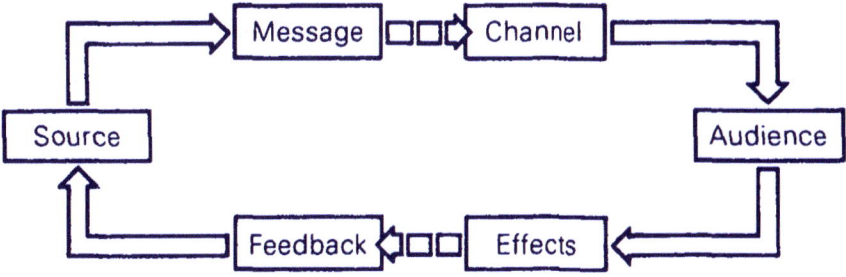

Let us isolate and examine some of the elements involved in a single act of communication — the conveying of a simple message.

Source

The source or sender of the message may be an individual, such as an extension worker, or a group, such as a Ministry committee. The immediate source may not necessarily be the person who suggested the message or who first began the communication process. For example, a health extension worker is the source when he communicates the message 'Wash your hands before preparing food.' He did not, however, think of this message in isolation from his colleagues. As part of a primary health campaign he is expected to tell local people how washing their hands can help prevent the spread of several common diseases. His role, therefore, is to follow a policy outlined by his superior officers and to use his special local knowledge to translate the policy into messages he can communicate effectively.

Message

The message contains the information or idea that is communicated. Its meaning can be expressed in different ways: language, written words, pictures — even by movement, gestures, or facial expressions.

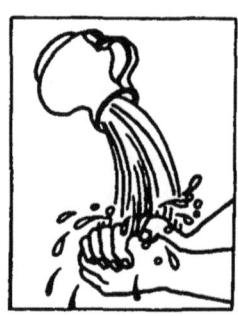

Pictures

Words

Demonstration

All these are symbols — they represent the information or idea that the source wishes to communicate. As a set of symbols, they have to be understood by both those sending it and those receiving it. This translation of an idea into a message to be communicated, and back again into an idea that has been received, is called 'encoding' and 'decoding'.

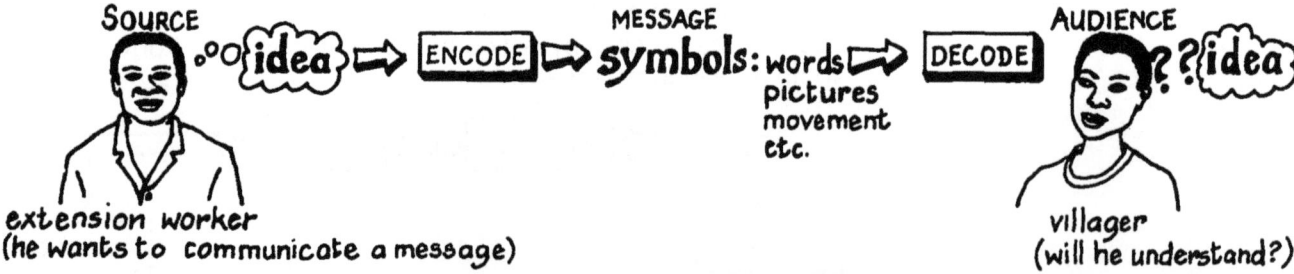

Many failures in communication happen because people make wrong assumptions about the way others will interpret their meaning.

Channel

The channel of communication is the means by which the message is communicated. Messages can travel from source to audience in many ways, for example:

- direct face-to-face exchange between source and audience, as in a discussion between extension worker and villagers;

- by way of a medium — this might be something as simple as a chalkboard or a written card, or it might involve mass media channels, such as radio or newspapers, which enable a single source to reach an audience of many.

Credibility

An important point about the means of communication is its credibility. This concerns the audience's opinion of the source of communication and the channels used. If these are considered trustworthy and competent, they are called 'credible' — and the more credible the source and channel, the more willing people are to listen to the message. Opinions of mass media are likely to be different from those of a personal approach, for example:

- The health extension worker is well-known and trusted by local villagers. When he visits them at home he is a welcome guest and can quietly talk over problems and give advice.

- The national newspaper carries weight as 'the voice of authority'; most readers believe what is presented to them in its printed pages.

Audience

The audience — the people who receive the message — is the most important single element in the communication process. The intended receiving group, or target audience, will hardly ever be 'the general public as a whole', and in fact it is much easier to communicate effectively if the target audience is limited to one particular kind of person.

Example:

'mothers'

is it all mothers, or can you specify mothers with babies under one year?

'farmers'

is it all farmers, or can you specify farmers owning land?

'villagers'

is it all villagers, or can you specify a group of village seniors?

It is essential to study the intended audience before making decisions about the message and the channel of communication. (See page 18 for information about target audience surveys.)

Communication Networks

Remember, too, that important lines of communication already exist in every community. For example, women may meet at the well and exchange news as they draw water; the local shopkeeper may have informed opinions and pass them on to all his customers. There may be links from child to child, or there may be some authority whose views are sought, such as a religious leader. This means that a message may be passed from the original (intended) receiver to many other receivers, spreading an idea far and wide. If a communicator examines local networks, he may be able to find certain very strong links, and if possible make use of them to help the transmission of his message.

Effects

A successful communication usually brings about change. Three main types of change can take place and usually in this order:

- a change in the audience's knowledge (what people know)
- a change in their attitudes (what people feel)
- a change in their behaviour (what people do).

Feedback

Feedback is a response to the message by the receiver, as observed and used by the communicator. Do not forget that communication should be a *two-way* process, since reactions to a message are in fact messages sent from the audience back to the source. Someone receiving a message will usually respond in some way, whether by words, a yawn, a shrug of the shoulders, etc. Many such responses are never seen or used; these cannot be called feedback, since feedback is the knowledge of results. The originator of the message can only be sure that the message has been both received and *understood* through the audience's response. He should use this response to modify further messages and revise materials as necessary.

Feedback can be grouped as follows:

- immediate, e.g. people's facial expressions may show if they have understood, and if the information is relevant;

- long-term, e.g. an ultimate change in behaviour;

- positive, where communication is satisfactory: the audience receive and understand the message and take the desired action;

- negative, where communication is not satisfactory and adjustments are needed.

It is easier to arrange feedback for some means of communication than for others. Conversation between individuals can provide direct, immediate feedback reasonably easily. Using media may mean there is less face-to-face contact between source and audience, and more formal arrangements may become necessary to collect feedback. When mass media channels are used, the source of the message is usually remote from the audience. It will therefore be difficult to collect feedback — this will need careful planning and organization. In general, the more that feedback is taken into account, the more effective will be the communication.

On page 9 we saw a simple diagram showing the idea of communication as a process. Here is another diagram that gives more information. 'Noise' refers here to anything that interferes with the message and stops it getting to the audience. This can happen in a number of ways, and at any point in the process.

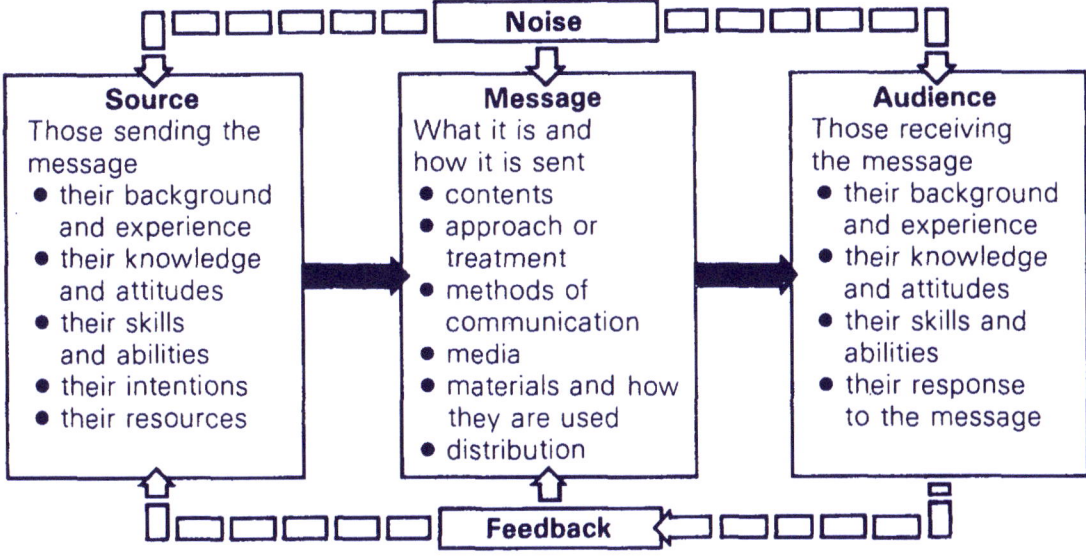

Communication as Common Understanding

Communication, however, does not work like an arrow shot from a bow to hit the target audience, or a dump truck unloading ideas from one person to another. It might be more constructive to think of it as a striving for common understanding. Both the author of a booklet and the person who reads it have their own fields of experience which influence them; both will have to make some effort to understand the other's point of view.

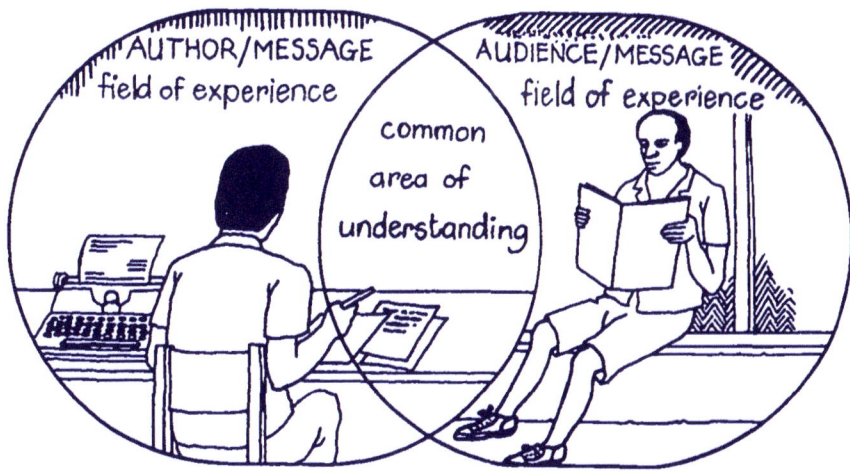

Only when a common area of understanding is established can we say that effective communication has taken place. Establishing understanding does not come easily or naturally — there are many barriers to be overcome. Communication is done for a purpose, and effort is required from both author and audience.

Problems in Communication

All too often communication is not effective. Despite the fact that more materials are being produced, more programmes broadcast, and greater amounts of information being spread to greater numbers of people year by year, campaigns still fail — diseases spread, crops are destroyed by pests, malnutrition continues. Failure in communication can be due to a number of different causes, and can occur at almost any stage in the communication process.

Source — the sender of a message nearly always has some responsibility for a failure in communication, perhaps through lack of care in research and planning, miscalculation of resources available, or lack of attention to feedback.

Message — this may fail because the information is incorrect, or because it is inappropriate to the receivers, e.g. something they have already rejected as unsuitable for their own conditions.

Channel — the choice of channel may be inappropriate, e.g. using the wrong language, using unfamiliar or difficult terminology, using a medium which does not reach the intended audience for technical or distribution reasons;

— credibility may be low, e.g. when the audience distrusts information from strangers.

Audience — information about the audience may be insufficient or inaccurate, so that the message is inappropriate or aimed at the wrong audience, as in these instances:

— the message is for too general an audience, therefore information is not precise enough to meet actual needs;

— the message runs counter to the audience's deep-rooted traditional beliefs and will be ignored;

— not enough explanation is given to convince the audience to change present well-tried ways of doing things;

— the audience cannot identify itself and its own problems in the information given, so will ignore it;

— the message is sent to an inappropriate but accessible audience, because poor planning of time and resources makes it too difficult to send to those in greater need;

— the audience is too poor or undernourished to carry out the desired changes;

— the audience is too concerned with quite a different problem to pay attention to the message.

Effects — the desired effects are not clearly defined, so the source cannot tell if communication has been effective.

Feedback — this may be ignored or not recognized;

— results may be missed if the time needed for long-term feedback is not appreciated;

— a producer's satisfaction with his work may tempt him to ignore negative feedback that would make him admit failure or revise his material.

Communication networks — sufficient attention may not be paid to these, e.g. the ideas of local opinion leaders may prevent an audience reacting favourably to a message.

Most of these communication problems can be prevented, or at least reduced, by preliminary research and careful planning. It is essential to have accurate background information about both the target audience and the proposed subject matter.

Examples

Readers may like to think further about the following failures in communication. What messages would have been more appropriate?

Problem: malnutrition, resulting particularly in ill-health among mothers and children and in a very high infant mortality rate.

Desired solution: an increased amount of protein in the diet of pregnant women, mothers and children. This could be obtained most cheaply and easily by encouraging women to keep chickens and use poultry meat and eggs to supplement the meals they prepare.

Main message: 'Eat chicken meat and eggs for better health.'

Feedback: people listened, smiled politely, but continued with their traditional diet.

Failure in communication: there is a deep-rooted local belief that if pregnant women eat eggs, their children will never grow any hair. The message conflicted with this belief and people therefore ignored it.

Problem: cattle disease has become widespread, local herds are weak and unhealthy, many animals are dying.

Desired solution: a programme of vaccination to protect cattle from disease.

Main message: 'Vaccination will protect your cattle.'

Feedback: people listened half-heartedly, and remained unwilling for their cattle to be vaccinated.

Failure in communication: people in the area are convinced that vaccination will weaken and kill their animals. They are therefore suspicious of the attempts to set up an immunization programme, and distrust the agricultural officers involved.

Chapter 1 Communication

> *Problem:* poor level of health among villagers; in particular, many diseases and parasites spread through the contact of unwashed hands with food.
>
> *Desired solution:* improved health through improved hand-washing practice (in the right way and at the right times).
>
> *Main message:* 'Wash your hands before touching food.'
>
> *Feedback:* people listened, but hand-washing did not increase significantly.
>
> *Failure in communication (1):* people accept disease as sent by gods, and are not really interested in hand-washing; they are more concerned with the failure of the maize crop and their worries about food for the year ahead.
>
> *Failure in communication (2):* people do not understand how dirty hands have any connection with the common diseases they suffer. Hand-washing therefore has no real meaning or importance; once the health workers had gone, people did not bother about it any longer.

> *Problem:* poor level of health among village children, in particular sickness and death due to diarrhoea.
>
> *Desired solution:* improved health through the correct treatment of diarrhoea by oral rehydration.
>
> *Main messages:* 'Diarrhoea is an illness'
> 'It may be dangerous or even fatal'
> 'One effective modern remedy exists (oral rehydration solution) that can overcome this danger.'
>
> *Feedback:* mothers did not adopt oral rehydration techniques.
>
> *Failure in communication:* people did not view diarrhoea as an illness, but as a normal occurrence in a child's life. It was not considered to be dangerous — after all, every young child suffered from it at some time or another and most continued to live in reasonably good health. Since diarrhoea was not seen as an illness, treatment was not thought necessary; people could see no purpose in preparing oral rehydration mixtures for their children to drink.

Chapter 2 Audience and Message

Chapter 1 dealt with communication as a whole process or experience, and introduced 'audience' and 'message' as elements within this whole. We saw that communication is more likely to be effective if the intended receiving group, or target audience, is limited to one particular kind of person and clearly specified. The message must be carefully designed with this group in mind. Let us move on now to look more closely at what you need to know about audience and message, and how you can find it out.

Take the example of an extension worker. He is based at a particular centre, but covers an area which includes several villages. He is a central figure in the education work and campaigns taking place in his area. Ideally he helps to create a two-way flow of information and ideas between his villages and the local and regional authorities. For this two-way flow to be effective, his educational programme must be carefully planned.

Finding Out about the Audience

First he must know *who* his audiences are. Every village in his area has its own local characteristics, and people have their own distinctive qualities too, as well as different knowledge, beliefs, professions, social status etc., giving rise to different needs and priorities. Before setting about his campaign, the extension worker might have to find out any of the following things about his audience:

Personal

— age, sex, marital status, state of health

— relevant interests, preferences

— relevant customs, folklore, beliefs, attitudes

— religious beliefs

Social and economic

— profession, employment

— status in family, social group, class or caste

— languages or dialect spoken, and at what level of comprehension

— standard of living

— relevant details of diet, hygiene

— day-to-day problems considered important (felt needs and priorities)

— existing communication networks, and how these can best be used to reach the target audience

Educational

— standard of education

— relevant knowledge, skills

— what media they already experience fairly often, and their media preferences

— what proportion can read without difficulty

- what proportion can recognize simple pictures or diagrams
- what proportion can understand abstract symbols
- what proportion can understand a series of pictures

Geographical
- area, climate, environment, water
- transport
- closeness to a main road, river, town (this also relates to experience of media)

About the subject matter concerned
- relevant traditions, habits, practices
- what they already know about the topic
- resources which are available locally

Visual details
- important visual aspects of the local culture which should be included in pictures (e.g. typical dress, style of housing, agricultural implements)

How to Find Out

TARGET AUDIENCE SURVEY

This information can best be discovered by a survey of the target audience. A survey is an instrument for increasing your understanding of the local people. If it is not possible to do a full field survey, obtain as much of the information as possible from existing knowledge (your own and your colleagues'), and also through observation and discussions *in the target area*. Assumptions about the target audience are just not accurate enough without a field visit to gain information on the spot.

Even an extension worker who himself came from the village will have distanced himself by his training, his mobility, his association with policy makers. He will not know as much about the village audience as he imagines; in addition, it will be difficult for him to take an unbiased view and keep his objectivity.

The subject of a campaign may well apply to all inhabitants of a village, or even of a district or province; but the approach used cannot be aimed in a general way at the whole population. For each topic to be communicated successfully to each target audience, a special set of conditions is needed, and the situation must be studied carefully in detail. Messages and materials must be designed for each specific audience. If specific audiences have not been considered before the background information is sought, this information will itself help to show which kinds of people should be grouped together. It will help production planning to specify each target audience as precisely as possible. Even an experienced media producer should make the effort to study each target audience before he begins production work.

Methods of collecting information:

- listening
- observing
- interviewing people and noting their responses; you may do this by filling in a questionnaire.

All three methods should be combined in the target audience study.

Steps in a Simple Survey

Preparations

1 Design a questionnaire. This is a list of questions that you ask in order to get the information you need about the audience. You might only need a few questions jotted down as a guide, to remind you of important points but not to be rigidly followed. Sometimes a carefully structured questionnaire will be needed, for example if you need to draw definite facts and figures from a number of interviews, or when many different people are doing the interviewing and would find it hard to be consistent about the way they do it.

A sample questionnaire is shown on pages 20 to 23. It is based on those used by officers in the Kenya Literacy Campaign, during workshop surveys. They wanted to find out as much as possible about learners' interests and needs, and to use this information to produce relevant teaching/learning materials and reading matter. The form therefore covers a number of different topics; it is carefully structured and quite long.

Some questions may be quite straightforward, for example:

Do you keep animals?

Other questions may be asked to cue or prompt — to encourage people to speak freely and in some depth, for example:

What kind of changes are you interested in for your farm?

Many people will wish to please by giving the answers they think you want to hear, so it may be important to ask a completely open question, and not to suggest in any way what sort of answer you expect or require.

Notice the difference between these two questions:

Would you like to be able to read letters and newspapers?

What kind of things would you like to be able to read?

Answers may not be accurate if the question leads people into replying in a certain way, or limits them to a 'yes' or 'no' answer, or indeed if the questioner suggests possible answers too quickly. People must be encouraged to think for themselves and to respond, when they are ready, from their own experience. Sometimes a follow-up question may be needed, to clarify information, to help someone answer a previous question, or even to check the accuracy of a previous answer, for example:

What is your age?

How do you calculate your age?

I'm not sure I understand you, please would you explain that to me again?

If the question

What do you dislike about the class?

proves difficult to answer, for example, it might be followed by:

Are there chairs or benches to sit on?

Is the class very crowded?

Chapter 2 Audience and Message

SAMPLE QUESTIONNAIRE - LITERACY MATERIALS SURVEY

LOCATION (area within 5 miles)

Name of place .... DATE ....

1. <u>What facilities are there here?</u>

        tick as appropriate   - primary school ....
                                          secondary school ....
                                            health centre ....
                                            other (specify) ....
                                            ....

PROBLEMS OF DAILY LIFE

2. <u>What sort of day-to-day problems do people have?</u>

        tick as appropriate   - water ....
                                              roads ....
                                              shopping centre ....
                                              employment ....
                                              other (specify) ....
                                            ....
                                            ....

RELIGIOUS GROUPS

3. <u>Which religious groups are active here?</u> ....

4. <u>Are any engaged in adult education?</u> (draw circles round the groups which are)

OCCUPATION

5. <u>What type of farming is done in this area?</u> ....

6. <u>What are the cash crops?</u> ....

7. <u>What do you grow on your own land?</u> ....
....

8. <u>Do you keep animals?</u> (if so, find out what) ....
....

9. <u>What kind of changes are you interested in for your farm?</u> (note the first two given)
....
....

10. <u>How much time do you spend on these things:</u> (read out list and tick appropriate box)

	much	little	none
cooking			
washing clothes			
gathering firewood			
fetching water			
looking after children			
cleaning			
building or maintaining the house			

BUSINESS

11. <u>What kind of business are you engaged in?</u>

> tick as appropriate – crafts ....
> transport ....
> retail ....
> hawking ....
> other (specify) ....
> ....

LEISURE TIME

12. <u>How do you use your leisure time?</u>

> tick as appropriate – ceremonies ....
> meeting with friends ....
> church ....
> choir ....
> dancing ....
> other (specify) ....
> ....

13. <u>Are there evenings of story-telling?</u> ....

14. <u>Do you listen to the radio?</u> ....

15. <u>What programmes do you like best?</u> ....

....

16. <u>Where do you listen to the radio?</u> ....

LANGUAGE

17. <u>Which language do you use</u> (a) <u>in the home?</u> ....
    (b) <u>in your business?</u> ....
    (c) <u>with Government officials?</u> ....

18. <u>Which written language is used here</u> (a) <u>for books?</u> ....
    (b) <u>for newspapers?</u> ....
    (c) <u>for announcements and notices?</u>

    ....

19. <u>Which language do you want to read?</u> ....

20. <u>Which language do you want to learn in?</u> ....

READING INTERESTS

21. <u>What kind of things would you like to be able to read?</u> (e.g. letters, shop signs etc., but let people think of their own answers)

....
....

WRITING INTERESTS

22. <u>Why do you want to write?</u> ....

....

23. Who will you write to? ....

24. What kinds of things will you write? (e.g. wills, etc. but let them speak for themselves) ....

PICTURES

25. Do you like pictures in your books? ....

26. What kinds of pictures do you like? ....
....

27. Do you like photographs or drawings better? ....
    (find out reasons if possible) ....

FACILITIES

28. Is there a place where you can read comfortably? ....
....

29. Do you find any problems in studying at home? (if so, find out what problems)
....
....

30. What other extension workers do you meet?

    tick as appropriate – health ....
                          agriculture ....
                          other (specify) ....
                          ....

31. How often do you see them? ....

32. How helpful are they? ....

LITERACY CLASS

33. What do you hope to gain by going to literacy class? ....
....
....

34. Where is the class held? ....

35. How far away is it? ....

36. What times of day are best for you? ....

37. How often do you go? ....

38. Are there any obstacles that prevent you going as often as you would like?
....

39. <u>Do you like the place where your class is held?</u> ....

                                    a lot it's all right a little not at all

       tick appropriate box -   ☐   ☐   ☐   ☐

40. <u>What do you like about the class?</u> ....

41. <u>What do you dislike about the class?</u> ....

42. <u>What books are there in class for you to read?</u> ....

....

PERSONAL

43. <u>What is your age?</u> .... <u>Sex</u> ....

44. <u>Are you married?</u> ....

45. <u>Do you have any physical disabilities?</u> ....

46. <u>Have you attended school?</u> ....
     (prompt as necessary to establish educational background)

47. <u>What is your social role?</u> (e.g. choir leader, club member, grandparent, etc. but let them speak for themselves) ....

48. <u>How long have you attended literacy classes?</u> ....

49. <u>How long will you continue?</u> ....

2 Try out your questionnaire on a few people to check that it can be clearly understood. The next step is to produce enough copies of the questionnaire; make sure that everyone using them understands exactly what each question means and what is intended.

3 Plan the method of survey.

4 Formalities — ask permission of anyone in authority who may be involved, e.g. village headman, local medical officer, teacher.

Carrying Out the Survey

A target audience may already be defined, for example mothers who attend a child-care clinic, literacy learners attending classes, villagers who sell food at the market. These groups have a known meeting-place; they can be visited at an arranged time and asked to help by answering questions about themselves and their interests.

If a whole village has to be explored and later divided into target groups, several people may be needed to help make the survey. One way is for each member of the survey team to take a different direction out from the village centre, and to stop at every other household. Here he may ask to speak to people belonging to a particular group, such as children attending school, mothers with babies, etc. or he may take a random sample. For example, during a workshop in Orissa, a survey was made to discover existing knowledge and practice regarding hand-washing. The target audience was divided into the following categories, and each group interviewed:

— adults (general)

— children up to 15 years old

— those who prepared food (women in the household)

— shopkeepers and stall holders selling food

— mothers with babies and young children (child care and child training).

Consider your approach and method of asking questions. It is important to make people feel at ease before you start. You may need to take time on introductions, or to explain what you are doing and the purpose of your questions. People may need to be reassured that the information is not being collected in order to levy taxes or impose bye-laws! Sometimes it will not be possible to collect accurate information. People may feel that information about things such as numbers of children, or possessions — radios, bicycles, etc. is private: they may not want to be counted or categorized. Be as encouraging as possible and try some follow-up questions, but be sensitive to your audience: where information is not willingly given, do not get upset — move on to another question, and try to supplement the answers given by observation and listening.

Analysing Results

You will return from the survey with a pile of completed questionnaires. The next task is to draw the facts you need from the information collected. Each survey, or perhaps each question asked, will dictate how this is done.

Look back at the sample questionnaire on pages 20 to 23. You will see that questions are asked in different ways. Question 11

Chapter 2 Audience and Message

can be used to show a straightforward way of analysing findings: group the responses together, and use arithmetic to get percentage figures related to the total number of respondents (people interviewed). In this example 25 people were interviewed and there are 25 completed questionnaires to be put together.

Question: *What kind of business are you engaged in?*

tick as appropriate

crafts _____   transport _____
retail _____   hawking _____
other (specify) _____

Responses				Percentage of total number of respondents
crafts	‖‖‖‖ ‖	=	7	28
transport	‖	=	2	8
retail	‖‖‖‖	=	5	20
hawking	‖‖‖‖ ‖	=	6	24
farming but no other business	‖‖‖‖ ‖‖‖‖	=	10	40

Question 22 is slightly more complicated, as some people had several reasons for wanting to write, but it can be analysed in the same way.

Question: *Why do you want to write?*

Responses	Percentage of total number of respondents
a. to do my accounts ‖‖‖‖ ‖‖‖‖ ‖‖‖‖	
b. to record local happenings ‖‖‖‖ ‖‖‖‖	
c. to write down family secrets ‖‖‖‖	
d. to keep in contact with relations ‖‖‖‖ ‖‖‖‖ ‖‖	

Readers may wish to work out percentages for themselves, and to explore which questions give the most satisfactory answers and results. More information about using questionnaires and analysing results is given under 'Pretesting' in Chapter 7.

Here are two other examples, involving the target audience's ability to recognize pictures. The first example deals with one picture at a time, takes the total number of respondents as a basis for calculation, and shows what percentage of them recognizes each picture. The purpose was to establish whether the pictures were clear enough to be understood by the target audience.

The second example analyses results in a different way. It does not separate the pictures tested, but treats them together as a single score. It takes a respondent's best possible total score as the basis for working out percentages. The purpose was to discover information about the audience, not the pictures. If you need to compare the results of different age groups of respondents quickly and easily, but do not need to identify each individual picture tested, this method will probably be more suitable.

Example 1[1]

In a survey conducted in Kenya to test recognition of drawings and understanding of visual symbols, people were asked what they saw in a number of pictures shown to them. A form was completed with the answers of each individual:

Place of interview	Sex Age	Tribe Education	Date
Card number	Response 1 = correct 2 = wrong 3 = don't know	If answer 'Don't Know' — ask 'What do you think it may be?' — record answer. If answer wrong — record what is said. Ask all who respond — 'What is it about the picture that makes you think it is a . . . state object . . . ?' — record answer.	

(up to maximum of 18 cards per interviewee)

The responses collected were grouped together and added up so that different groups could be compared. Part of the resulting table is shown on page 29.

Some information is immediately clear, for example:

No.19 snake — nearly 100 per cent recognition.

Other facts can be seen when the different groups are compared and percentages worked out, for example with No.36, maize:

Educational level	Number of correct answers	Total number of respondents	Percentage of respondents giving correct answers
Standard 1	56	76	74
2	66	86	77
3	99	111	89
4	130	147	88
5	110	113	97
6	109	110	99
7	60	61	98
Non-literates	74	131	56
Rural non-literates	61	199	31

Chapter 2 Audience and Message

The figures show that recognition appeared to improve with education. Of course you will not usually need to write out all the information given in the two middle columns; you can set out the final results more clearly with just the first and last columns.

Example 2[2]

This is designed to show a way of working out percentage scores from interview responses. A target audience survey has taken place, with research into the audience's ability to recognize pictures. Interviews were held, and the respondents were asked to recognize household items shown to them in pictures. Their scores are shown below. Note that respondents were classified by area (A, B, C, or D), by literacy and by age. Note also that in each of these sub-groups, seven interviews were conducted. A perfect score for each interview was four points; points were lost as people failed to recognize items in the pictures.

Scores for recognition of household items				
	Area A	Area B	Area C	Area D
Literate Age 15-50	4,4,4,4, 3,4,3	4,4,3,3, 4,3,4	4,4,3,2, 3,4,3	3,2,4,4, 3,4,3
Age 51 plus	3,3,3,4, 4,3,3,	4,3,2,4, 3,2,4,	4,4,3,4, 2,4,2	3,3,2,3, 2,3,2
Non-literate Age 15-50	3,2,3,2, 4,2,3	3,2,3,4, 2,3,2	3,4,2,3, 2,2,2,	3,2,2,2, 3,2,2,
Age 51 plus	3,2,3,2, 3,2,3	3,2,3,4, 3,2,2	3,4,2,3, 3,2,2	2,1,1,1, 3,2,2,

After scores from the interviews have been written out like this, add up the scores in each sub-group. For example, in the top left sub-group, the scores (4,4,4,4,3,4,3) are added up to give a total of 26. After this has been done for each sub-group, add up the total across and down for each column and line. Check your answers with the following results.

Sub-group totals for area, literacy, and age					
	Area A	Area B	Area C	Area D	Total score
Literate Age 15-50	26	25	23	23	97
Age 51 plus	23	22	23	18	86
Overall literate	49	47	46	41	183
Non-literate Age 15-50	19	19	18	16	72
Age 51 plus	18	19	19	11	67
Overall non-literate	37	38	37	27	139
Total score	86	85	83	68	322

Next, note the maximum possible number of points that could have been achieved in each sub-group, and work out the percentage score. For each sub-group here, there were seven interviews with a possible four points perfect score for each interview. So in the top left sub-group, for example, the number scored is 26, while the maximum possible score is 28. Twenty-six is 93 per cent of 28, so the percentage score for the sub-group is 93 per cent. Answers can be checked with the results in the next table.

Percentage scores for picture recognition grouped by area, literacy and age					
	Area A	Area B	Area C	Area D	Overall
Literate					
Age 15-50	93	89	82	82	87
Age 51 plus	82	79	82	64	77
Overall literate	87	84	82	73	82
Non-literate					
Age 15-50	68	68	64	57	64
Age 51 plus	64	68	68	39	60
Overall non-literate	66	68	66	48	62
Overall (literacy, age)	77	76	74	61	72

This study shows that literates performed better than non-literates (82 to 62 per cent), that older people had slightly lower scores whether they were literate or not, and that Area D had lower scores than the other areas. Note that other percentages could be worked out from the numbers in this table, and other variables (such as content and art style) could be broken down similarly.

More information about using questionnaires and analysing results is given under 'Pretesting' in Chapter 7.

Preparing the Message

Finding out about the intended audience is an important part of your planning; so, also, is careful preparation of the message you wish to communicate to them. Let us return to the example of the extension worker. He has conducted a simple survey and discovered many useful facts about his audience. He many have been surprised by the amount they knew already, and by some of their views on his own ideas for improvements in their villages.

He must now think about *what* his message will be. Although the general pattern of his work will be directed by the policy of the health authorities, and by the current campaigns taking place, on a local basis he should be able to identify:

- the needs that require most urgent attention
- the change(s) in behaviour that would help the audience to meet these needs.

Understanding of Visual Symbols in Kenya: Recognition of Pictures

Drawing	Answers given	Totals			Standard				Totals			Standard			Totals		No Education	
		All	Pre-test & Part 1	Rural	1	2	3	4	Standard 1 - 4	Rural	5	6	7	Fm.1	Standard 5 - Fm 1	Rural	Pre-test & Part 1	Rural
No. 36 Maize	Correct	863	707	156	56	66	99	130	351	52	110	109	60	3	282	43	74	61
	Incorrect	168	86	82	7	12	10	15	44	14	3	1	1	—	5	4	37	64
	Don't know	129	45	84	13	8	2	2	25	10	—	—	—	—	—	—	20	74
	Total	1160	838	222	76	86	111	147	420	76	113	110	61	3	287	47	131	199
No.19 Snake	Correct	303	267	36	23	35	45	29	132	4	41	47	29	2	119	2	16	30
	Incorrect	—	—	—	—	—	—	—	—	—	—	—	—	—	—	—	—	—
	Don't know	4	1	3	1	—	—	—	1	—	—	—	—	—	—	—	—	3
	Total	307	268	39	24	35	45	29	133	4	41	47	29	2	119	2	16	33

Deciding on the Topic

This may be directed from higher authority, or may arise from the extension worker's own assessment of local priorities. His choice of topic should reflect:

- the particular problems of the audience
- their attitude to these problems
- the resources available for change.

It would be difficult, for instance, to persuade villagers to build good latrines if they:

— consider latrines unhygienic;

— prefer to use open fields;

— cannot afford to build latrines, and feel anyway that there are more urgent priorities for the little money they have;

— do not have access to suitable building materials;

— cannot dig pits because the ground is too rocky.

He needs to make a realistic appraisal of the audience needs, attitudes and resources discovered during the target audience survey.

Studying the Topic

The topic chosen must be broken down into its different parts — each part is itself a message. Let us look again at the topic of hand-washing. This may seem to be a simple and straightforward subject, but in fact the information breaks down into a number of smaller subjects:

— how diseases spread, for example by germs

— the connection between germs and dirt (dirty hands)

— how hand-washing prevents the spread of disease

— how hands should be washed

— when hands should be washed.

Some of these may be more complex than they first appear. Take, for example, the idea of 'germs'. *The village audience cannot see or feel germs*: they have never seen a microscope, they are not familiar with enlarged scale illustrations, diagrams, anatomical drawings or cross-sections. It will need a very careful and well-illustrated explanation to convince them that germs cause disease — the basic reason for persuading people to wash their hands.

Each message must be carefully thought out in relation to those who will receive it. The information gathered during the survey will provide valuable guidelines, for example:

- what do the target audience know already?

There is no point in teaching them something they already know — they will feel insulted or lose interest.

- what *are* their feelings (emotions) about the subject matter?

A farmer may understand the reasons for using better quality seed, but he is not likely to make any change if he feels that the campaign for better seed is really designed to make profits for the seed merchants.

- why are they not already doing what you want them to do?
There may be a good reason why your method does not work in this area.

- what is economically possible for them?
If you are recommending the use of soap, special foods, contraceptive pills, etc., are these things available in the area, and do people have enough money to buy them? Even if a farmer fully understands the need for change and is willing to adopt new methods, he must have the necessary resources to take action, for example:
— if fertilizer is expensive, and no provision is made for getting loans at a reasonable rate of interest — no action;
— if the money is there, but no fertilizer is available when the farmer needs to use it — no action.

Preliminary research will give answers to questions such as these, and so will influence the message chosen and the way in which it is communicated.

Check that the proposed message is completely accurate — that enough information is given and that all details are correct. The extension worker might wish to check his material with a specialist such as a doctor or agriculturalist. He might be able to find someone who has already done what is being proposed, such as an innovative member of the target group, and ask about their experiences. It is better to check several times than to risk losing credibility because the audience find that they have been given incorrect information; once their trust and confidence is lost, it will be difficult to regain.

Example

> A campaign was organised to persuade farmers to use fertilizer and thus increase ground-nut production.
>
> *Main message:* 'Use fertilizer to increase your ground-nut production.'
>
> Not enough information was given about the reason for using fertilizer, the right kind to use, and the method of applying it; farmers were left to read the instructions on the fertilizer bags for themselves.
>
> *Result:* Several farmers used too much fertilizer and seriously damaged their crops. Not only did they lose money, but they suspected the whole campaign, feeling that it was a political plot to ruin them.

Chapter 3

Aims and Objectives

The extension worker has researched both his audience and the relevant subject matter. He has divided the intended audience into appropriate target groups; he has chosen a topic, broken it into its component parts and thought about the appropriate messages. He can answer the first two key questions on page 5 and is ready to ask himself the third question:

What results or reactions are desired as a result of your message?

Aims

He almost certainly had some kind of purpose in mind from the start, probably something quite vague, such as 'improving the villagers' health' or 'improving public cleanliness (sanitation) in the village'. After his research, he may now be able to outline — still in general terms — the purpose of the task in hand, and the way in which his messages would help to achieve this purpose, perhaps like this: 'to improve sanitation in the village by encouraging people to build latrines and use them properly'. These general ideas of intention and purpose are called *aims*. They are a necessary first stage, but are not very helpful when it comes to making specific decisions or trying to discover whether a communication has achieved its purpose.

Objectives

For this it is necessary to take a closer look at the results required, and to specify *exactly* what the audience should know, feel or do as the outcome of the communication. These specific statements of intention are called *objectives*. They provide a basis for planning, developing and presenting media, and are also essential when carrying out testing and evaluation. Results achieved are linked with the original objectives, to see how they compare with what was intended.

Objectives are most often concerned with actions — what people should be able to do as a result of the communication. When writing objectives, try to use active, definite verbs so that the resulting actions can be clearly seen or heard.

Verbs like 'recite' are good; verbs like 'believe' are much harder to test — how do you know if someone really believes what you tell him?

```
know        —— no
recite      —— yes
explain     —— yes
understand  —— no
appreciate  —— no
construct   —— yes
list        —— yes
believe     —— no
```

Example

Let us look at the health extension worker and his messages about latrines, and see how he develops useful objectives.

Aims

- To upgrade the general level of health in the village by improving sanitation and thus reducing the spread of many common diseases;

- to encourage people to build latrines and use them properly and not to defecate near houses or water supplies.

These aims are not in themselves precise enough to help make decisions on the production of materials. Remember how research can help.

- The target audience can be divided into appropriate groups, for example:

— adults

— children up to 15 years old

— mothers looking after babies and young children.

● The subject matter can be broken down into its component parts, for example:

— how diseases spread (germs and parasites);

— connection between germs, parasites (worms, eggs etc.) and dirt (faeces, dirty hands, contaminated food and water) and how dirt reaches people's mouths and spreads infection;

— how cleanliness and good sanitation (including use of latrines) can reduce the spread of disease;

— how to build a good, safe latrine;

— how to use the latrine.

Objectives

By the end of his campaign, the extension worker wants his entire audience to be able to do the following:

1 Explain how dirt can spread disease;

2 List four methods of using good sanitation to prevent the spread of disease;

3 Defecate away from houses and water supplies.

In addition, he wants at least five out of the twenty families in the village to:

4 build a good, safe latrine and to

5 use the latrine properly and keep it clean.

All these objectives are precise enough to be tested or evaluated in some way, whether by observation or by asking questions. The communicator can find out whether his campaign has been effective, or whether people are not convinced of the benefits of good sanitation and still keep to their previous habits.

As well as being specific, objectives must be attainable and worth-while; that is, it must be possible for the audience to achieve the objectives, and the audience must feel that it is worth doing so. This last point is sometimes overlooked: technically sound projects may fail if the importance of human attitudes, feelings and emotions is not appreciated. (Remember the audience characteristics discussed on page 17). A member of the target audience, whether villager, small farmer, peasant, estate worker or anyone else, is not just waiting there to be shaped and guided. He is living his own life and running his own affairs; he is intelligent even if uneducated and has beliefs, opinions and feelings; he knows his environment and his problems better than an outsider. It may be difficult to persuade him to leave the security of his existing routine for change, risk and extra work, in order to achieve some promised benefit — a benefit which may not be clear to him, and may even be questionable from his point of view.

Motivation

Information alone will not cause people to change their normal habits, although it may give good reasons why they should. Attitudes or feelings are involved — they must be convinced about the value of change to themselves. This feeling of conviction and positive wish to change is known as *motivation*.

Chapter 3 Aims and Objectives

All sorts of things can motivate people — a wish to improve life for their families, a fear of sickness or death, and so on.

The target audience survey will provide information about what people think they need and want, and will thus throw light on the factors which are likely to motivate them towards change. Of course it is not just a matter of asking what people want and getting a simple answer. Few of us can accurately identify and explain our own problems; people may need help in exploring their needs, discovering priorities and deciding what solutions are possible.

Finally, the communicator should also be able to think beyond his immediate objectives, and be aware of how they relate to the general situation, or even to broader aspects of national development. It may be that solving a problem turns out to be only one step along a very long and difficult path, for example:

- if farmers produce more ground-nuts,
— who will buy them?
— how will they be taken to market?

- if a river is diverted to irrigate the fields,
— what should be done about the fishermen who depend on the river for their livelihood?

It may even be that solving one problem in a particular way creates new and worse problems, for example:

- a new plough is introduced to increase farmers' productivity, but it digs so deeply into the light local soil that large areas dry out and are blown away by the winds;

- a river is dammed to provide a better water supply, but the resulting standing water brings devastating new diseases to the area.

Summary

Using general aims and his analysis of audience needs and priorities as a guide, the extension worker has been able to do the following:

- identify the target audience: divide it into specific, tightly defined groups;

- identify the topic: break it into component parts and select appropriate messages;

- define objectives (by marrying together the audience groups and the messages, and specifying the required results).

He is now clear about the exact nature of the task involved, and of the expected results. He can go on to consider:

- the way in which his message should be presented, and how treatment should be matched to the expected audience reaction, for example:
— whether it will inform, direct, forbid, warn, persuade;
— how it will seek to motivate;

- how it can be transmitted to his audience:
— what media and materials are most suitable;
— how material will be used (e.g. by a teacher or extension worker, or by individuals on their own).

Section B

Making Choices – Methods and Media

Chapter 4

The Range of Possibilities

Having answered the questions *WHO?* and *WHAT?* the next step is to consider the question *HOW?* What is the best way of getting the message across to the audience? To answer this, you need to know something about the range of possibilities available, in terms of:

- the method of communication
- the kind of approach made to the audience, the kind of session or campaign that is planned, the way in which any supporting media are used;

- the media
- the kinds of materials that could support the planned approach.

Factors in Making a Choice

There is a wide range of techniques from which to select the best method and the most appropriate medium. It is usually possible to work out a number of different ways of communicating messages in order to achieve a stated objective. You will need to make choices, and the more relevant information available, the better informed will be the choice.

Consider and compare the different possibilities in relation to the audience and to the message, for example:

- which are familiar or acceptable to the audience?
- which are available to them?
- is one common language suitable for all audience groups?
- are people to be shown how to do something practical?

Two other important factors are the resources available and the constraints which may apply, for example:

- how much money is available?
- what personnel are available? (production staff, field workers etc.)
- how much time is there?
- what distribution channels are available? (postal services, radio networks, electricity, transport etc.)
- do personnel have to be trained to use the materials?

It will help to make a list of all available methods and media, and to consider each one in terms of your audience, message, resources and constraints.

A Multi-Media Approach

It will probably be most effective to combine several different methods and media — a multi-media approach. This will give a number of advantages:

- different methods and media have different strengths and weaknesses: a combination can use the strong points of each, and use them for the particular subjects or aspects they are best suited to; (characteristics of different media are discussed in the next chapter;)

Chapter 4 The Range of Possibilities

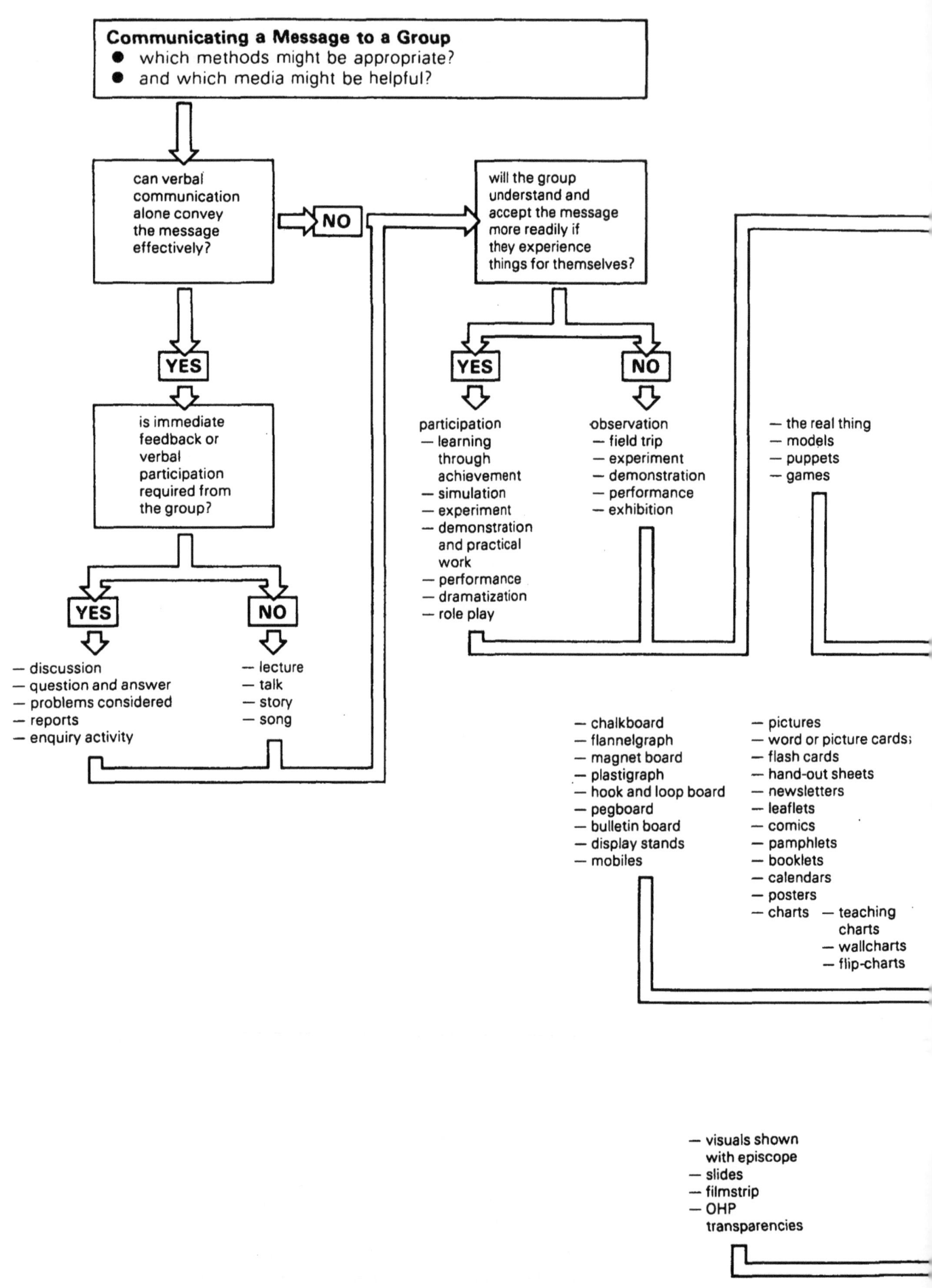

Chapter 4 The Range of Possibilities

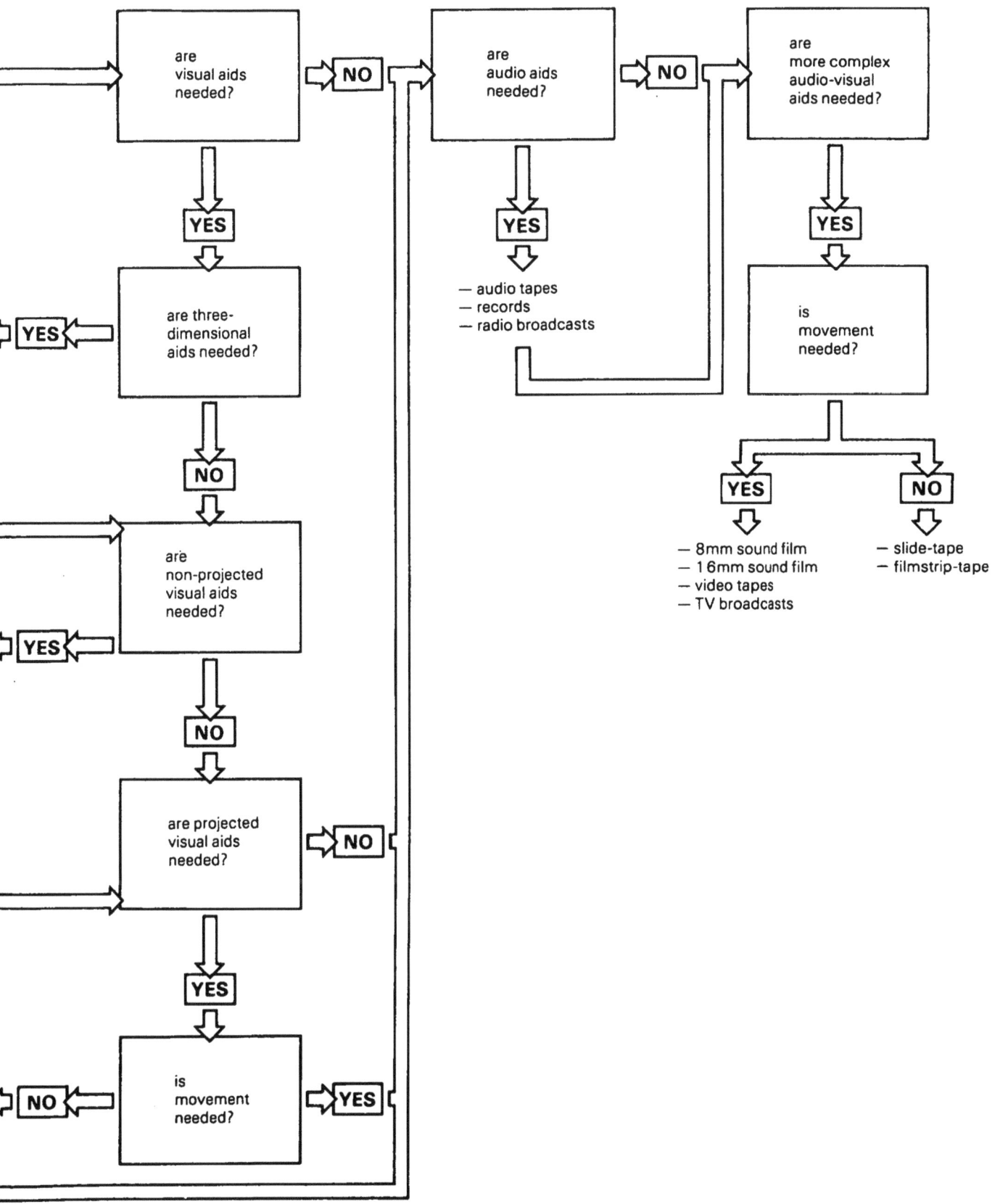

Chapter 4 The Range of Possibilities

- people vary in the approaches and materials they respond to: a range will therefore have a wider appeal, as well as overcoming differences in education, literacy level etc;

- distribution methods differ: a range of media makes it easier to overcome problems and reach a larger number of people; e.g. where radio reception is poor, distributing recorded cassettes or printed material may be a better solution;

- different methods and media can be used to reinforce each other by presenting information in different ways, by emphasizing different aspects, and by encouraging different responses from the audience.

Of course your choices will normally be predetermined by factors such as existing equipment and technical staff, or by criteria such as:

— which is the cheapest?

— which reaches the largest audience?

Methods and Media — an Overall View

For the moment, however, let us imagine a fresh starting point, and look at an overall view of the range of methods and media. The possibilities are confusing by their number and variety, and it is difficult to work your way through them in order to narrow down the field and select the most appropriate. The diagram on page 38 presents such an overview. It offers a way of taking the different choices step by step, and guides you through them with a straightforward question and YES/NO answer routine. Communication methods are grouped on the left-hand side of the page, followed by the different aids or media — beginning with the simplest, and reaching the most expensive and complex on the right-hand side. Printed material is included in the list of non-projected visual aids.

Chapter 5   Narrowing the Field

Each stage of communication so far has involved making choices and narrowing the field — for example, the need to select specific audience groups and pick out the most appropriate messages. Now we can do the same with methods and media. Having seen the overall picture shown in the diagram, we should look more closely at the techniques most likely to be appropriate to education and development.

Communication Methods

The audience receives information through its senses: sight, hearing, touch, taste and smell. Sight and hearing are the dominant senses and most communication appeals primarily to these. It is often easier to understand and remember by seeing than by hearing: 'a picture is worth a thousand words' — if it is the right picture. In some places, however, there may be a strong oral tradition such as story-telling, and here pictures will have less impact. It is usually even better if the audience can actually do something for themselves. This gives the greatest possible amount of active participation — the most concrete experience possible. (and therefore the most easily understood and remembered).

Best of all is a combination of different methods and media, so that people use different senses to take in information and are actively involved in the communication process. Methods which are often used include: learning through achievement, arranging a 'simulated' experience (models, games, drama, puppetry), story-telling, song, discussion, field trips, demonstrations, talks, lectures and exhibitions. We look briefly at these, and list some advantages and disadvantages that occurred to the author. Of course advantages and disadvantages depend on the circumstances and you will think of others for yourself.

Learning through Achievement

Learning by doing involves concrete experience and active participation. The learner applies his existing knowledge, adapts it in the light of new information and experience and takes responsibility for the results of his actions. It is probably the most effective way of understanding and remembering, particularly if it is guided so that the learner can achieve enough to encourage further progress.

Advantages:
— it is good for learning practical skills;
— it is good for understanding something in depth.

Disadvantages:
— it may be difficult or expensive to organize;
— it may need a lot of time (from both communicator and learner);
— it may not be practicable for more than a few people at once;
— it may take a long time to achieve results.

A straightforward example is the mastering of a practical skill such as carpentry, where people can work as apprentices and learn on the job. Achieving a successful result encourages them

Chapter 5 Narrowing the Field

to repeat the exercise, remember what they have done and go on to try more advanced work.

Here is a more complex example. The co-operative extension worker has encouraged some women to start a small cottage industry silk-screening cloth. They use existing skills, but are helped to adapt them to a larger scale operation and use them in a commercial way — to print the first batch of cloth and sell it successfully. The success of these first steps gives the women a feeling of achievement and confidence. They believe in what they have learned, and are encouraged to continue building up their business and learning to improve their skills.

Simulation

Learning by doing can take place in a real life situation, but also through a 'simulated' experience. Here, circumstances are arranged so that people can take decisions and try things out freely before being exposed to real life, where they would have to face the consequences of those decisions. Examples of simulation are the use of models and games. Drama, role-playing and puppetry can also simulate reality, through the acting of different roles. With simulation, the learners need to make a mental jump back to reality; they must use imagination to relate what they are seeing or doing to real life. It may sometimes be an advantage for people to be able to distance themselves from the subjects they are considering. For example, they can freely discuss a problem presented in a drama, without the possible embarrassment of admitting it to be their own problem.

Simulation: Models

A model is a manageable representation of the real thing, useful when it is not practical or appropriate to make use of a real object.

Advantages:
— models are good for explaining clearly how things work; three-dimensional models are more easily recognized than two-dimensional pictures (though scale must be correct throughout, or it may confuse);
— they can involve the sense of touch as well as sight (this increases familiarity and confidence);
— they can include moving parts (good for learning about equipment etc.);
— models are good for simplifying complex principles — they can be designed to stress important points and to leave out unnecessary and distracting details;
— they can be good for trying out alternatives and contrasting different arrangements.

Disadvantages:
— they may be difficult to design and expensive to make, particularly if they must be realistic and accurate;
— they must be able to withstand wear and tear;
— they may have very limited use within a communication session;
— they may be difficult to transport and to store when not in use.

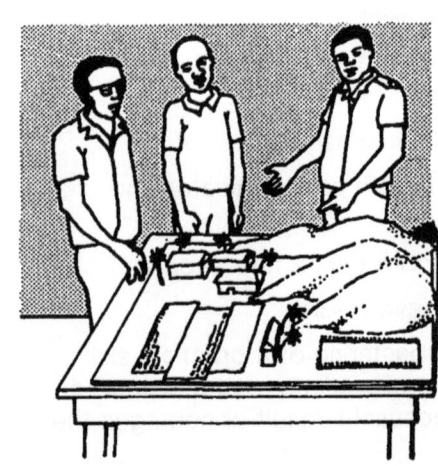

There are many different ways of making and using models: the model farm shows a good way of explaining things and exploring new ideas. It is three-dimensional, it can be seen and touched, looked at from all angles, dismantled and put together in different ways. The farmers find it clearer than paper maps; it involves plenty of audience participation and stimulates discussion.

Simulation: Games

Many people enjoy playing games. Games can be used as another way of helping people to think things out for themselves and to involve themselves actively in learning. A game based on an activity such as selling produce or developing a poultry unit can include most of the choices and problems faced in real life. There are many different kinds of game; some of them need materials, for example dice, chance or opportunity cards, mock money or other resources (in a central bank or held by players) and counters or pieces to move along a board.

Advantages:
— games stimulate interest and encourage active participation; popular local or traditional games can be adapted and games can be made competitive between individuals or groups;
— they are useful where different actions and outcomes must be explored;
— they are good for helping recognition and familiarity, and building up confidence (e.g. literacy games with word cards).

Disadvantages:
— it may be difficult to design a game that works well as a game and also achieves its communication objectives.

Simulation: Drama

Drama also gives people the opportunity to look at possible courses of action before committing themselves. Those watching the performance will receive information in an entertaining form and learn through both emotion and reason. The messages will be still more effective if spectators are called upon to react to the ideas presented and to give their opinions in discussion. People can even take part in drama or role-playing exercises in themselves. Role-playing can bring insight into different attitudes and opinions. Community drama encourages active participation by local people: they can choose the topic they feel is important, and therefore be actively involved with the message. They can express their views, analyse problems and identify solutions.

Advantages:
— it is good for studying a conflict or problem in depth;
— it is good for exploring the advantages and disadvantages of different ideas;
— it can have considerable impact.

Disadvantages:
— it may be difficult to organize;
— community drama will need time and effort from all participating.

Simulation: Puppetry

A puppet show is an entertaining way of communicating the essence of a situation or problem, although the realism of drama or role-playing is lost. As with drama, it can be organized as a community activity and can draw on a wide range of local talents, such as carpentry, sewing and acting.

Advantages:
— it is good for ideas that can be expressed in an exaggerated form, to leave the audience with a strong clear message;
— it is appropriate for presenting controversial ideas (e.g. family planning); the audience will accept unusual situations or controversial statements more readily than from live actors;
— it may relate well to popular traditional entertainment such as shadow puppets;
— it can have considerable impact;
— puppets can be made from all sorts of objects and materials;
— they can be simple yet effective.

Disadvantages:
— it cannot express subtle ideas or complex messages;
— it may be difficult to organize;
— a community production will need time and effort from all participating;
— it may not be acceptable to adult audiences.

Story-telling

A narrator may be involved in drama or puppetry, to tell the story, point out details, explain underlying themes, sum up, lead discussion and so on. Traditional story-tellers can help to convey educational messages, and stories passing from person to person could be an effective way of using communication networks.

Advantages:
— a cheap method;
— it can have impact;
— its relationship to traditional forms of communication can encourage understanding and acceptance of the message.

Disadvantages:
— it cannot convey complex information.

Song

This is a lively and attractive way of communicating, which can draw on local talent and express local views. It can be used in a similar way to story-telling and a simple song can also encourage the audience to join in and sing themselves, which will help them to understand and remember the message.

Advantages:
— a cheap method;
— it can have impact;
— its relationship to traditional forms of communication can encourage understanding and acceptance of the message;
— it is good for reinforcing a message through repetition (simple messages and catchy tunes stick in the memory).

Disadvantages:
— it cannot convey complex information; its message is clear and direct.

Discussion

Discussion is important because it involves two-way communication. It encourages people to express their views and help make decisions, and enables individuals to take part in the affairs of the community. People are more willing to listen to a message and to accept it if they have had a chance to think it over, argue about it, and even to help decide what it should say.

Discussion can strengthen all the other communication methods described if it is well-organized, so that everyone has a chance to speak if they wish. It can draw information from other kinds of activity, identify key points, involve the audience more deeply and encourage them to go further, make decisions, take action and so on. Discussion can make communication more effective, whether used before, during or after some other kind of learning session or community presentation.

Advantages:
— it is a valuable way of getting information about the audience and their views;
— it can easily be combined with other methods.

Disadvantages:
— a good chairman is needed to restrain over-talkative people and encourage those who are shy.

Field Trip

People can gain knowledge and understanding from seeing the real thing in action, whether it is a successful clinic, a new training method, or a technician's workshop.

Advantages:
— it can have considerable impact, if people are properly briefed.

Disadvantages:
— it can be difficult or expensive to organize.

Here is an example. Villagers have come to visit a successful co-operative, and to see its cannery, storage system and consumer shop. The main purpose of the visit is to inform and motivate. By seeing what can be achieved, and by hearing about it directly from the people concerned, the visitors will learn new methods, and should feel inspired to try to make some improvements of their own.

Demonstration

The audience is shown how something is done and also what the immediate results are. It is important that a demonstration is made clearly and at the right pace for the audience. It is usually best followed by practical work so that people can try things out for themselves.

Advantages:
— it is good for showing practical skills and set procedures;
— it is useful for showing immediate results, cause and effect (e.g. first aid, using machinery, food recipes);
— it can have impact.

Disadvantages:
— it may be difficult to ensure that everyone can see the demonstration;
— it may be difficult to organize;
— it may be difficult to link a demonstration with other methods;
— you might have to choose an audience with the necessary skills to copy what is demonstrated.

Talk or Lecture

A talk may be something quite informal, perhaps to a small group of learners, or even to a few people in their own home. A lecture usually means something more formal or academic, with the speaker distanced from his audience by status, surroundings or the size of the audience. Talking to people is a time-honoured way of communicating but it will not always be effective; a session where one person talks can easily be too long, too dull, too complicated and too abstract, or inappropriate in some other way. Remember that the more senses the audience uses, the more likely it is to understand and to remember. The speaker can combine a range of communication methods to involve people in different activities and enable them to use as many of their senses as possible. He can talk to them, show them things and give them things to do themselves. He can draw on a variety of resources and aids (chalkboard, books, episcope etc.) to help make his session clear, informative and interesting.

(a)

(b)

Advantages:
— a cheap method when talking to a large audience;
— much information can be given in a short time.

Disadvantages:
— it is often taken for granted, and not enough care taken to make it interesting and effective.

Both demonstrations and lectures can inform and explain; the two methods can be complementary. For example, in picture (a) an instructor is showing a group of learners how to prepare an improved kind of poultry food; he is showing them by actually doing the mixing, stage by stage. In picture (b) he is explaining the benefits of the improved diet in a lecture.

Exhibition

Exhibitions can make a powerful impression by using a variety of resources and different communication techniques (models, charts, books, games, demonstrations etc.). Learners can set up and run an exhibition themselves, so there can be a strong element of learner participation.

Advantages:
— it is good for public relations work;
— it is good for attracting attention, alerting the public;
— it can offer a wide range of information.

Disadvantages:
— it may be expensive to achieve the desired standard of impact;
— it may be difficult to organize.

Chapter 5 Narrowing the Field

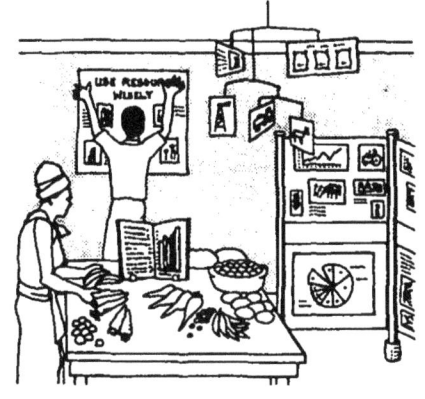

The picture shows an exhibition set up by a local co-operative. It is intended to inform and encourage existing members, and to show other visitors what a good co-operative can offer its members (public relations). The different elements shown in the picture include:

— display of produce grown;

— table-top display board: diagram of money made by selling produce;

— display stands: diagrams showing membership numbers, annual accounts, and photographs of co-op activities;

— mobile: picture cards to show things people can achieve through co-operation (water supply, cattle dip, better transport etc.);

— posters and charts.

Low-Cost Media

Most of the communication methods discussed rely upon the use of media to some extent; sometimes methods and media are related so closely that they can scarcely be separated. When talking about media it is also easy to confuse the equipment used with the materials produced. Let us concentrate on the materials you can produce yourself, and think of them as aids which can give valuable support to all the methods just described. For example, they can be used to:

— give extra information

— explain something clearly or in more detail

— illustrate a theory

— stimulate discussion or other learner activity

— attract attention, arouse interest

— help people to remember information

— sum up, reinforce information.

They can also be used as introductory material beforehand or as reinforcement afterwards. Aids may be designed for one particular occasion, or used several times, or stored and re-used on a regular basis. Learners can often be involved in the planning and production of materials.

Community Involvement

It will often be unrealistic to expect much participation from local people (who, after all, have many other demands on their time and energy); activities such as drama and puppetry may be too complex and time-consuming to arrange as a community production. Nevertheless, people will usually feel more interested in something and committed to its success if they have had even a small part in its development. If the target audience is consulted over the selection of topics or involved with the production of media, the message is more likely to be understood and accepted. The picture shows a booklet being pretested. Pretesting gives you another chance to check that the contents and presentation are on the right lines before spending money on the final production or multiple copies (see page 79 for information on pretesting).

Chapter 5 Narrowing the Field

Let us look at those media which are most likely to be reasonably accessible at a relatively low cost. They include sound recordings and a wide variety of visual aids: projected aids (photographic slides, overhead projector transparencies etc.), three-dimensional aids (models, games, puppets), printed material (posters, charts, calendars, flip-charts, flannelgraph pieces, word or picture cards, hand-out sheets, newsletters, leaflets, pamphlets, comics and booklets).

Sound Recordings

Both recordings on tape and material broadcast over the radio network can be interesting and relevant to the audience and can be useful resources for extension workers or instructors. While recordings on tape can be used freely as required during sessions, radio programmes must be heard when they are broadcast — unless, of course, they can be recorded.

(a)

Advantages:
— they are good for reinforcement (e.g. songs and dramatic sketches);
— they are good for information and motivation (e.g. messages from members of one co-op to another, or from important local people or public figures);
— they can capture items of local interest (e.g. interviews, case studies);
— they can encourage lively learner participation (e.g. songs and interviews);
— the equipment needed is relatively easy to obtain and to use.

Disadvantages:
— there is no tangible material that the audience can touch, look at, inspect carefully at their own speed, or keep and look at again.

(b)

(c)

Picture (a) shows an interview being recorded — a local farmer is talking about his work. In picture (b) villagers have been invited to a local radio studio to take part in a programme. Picture (c) shows a group of villagers listening to some recordings with an extension worker. The extension worker can monitor the listening group and note reactions to the material; this will help him to organize discussion etc. linked to the recorded material.

Visual Aids

Visual aids employ the powerful sense of sight. They can be extremely useful in training, extension work, project planning, sales and marketing drives, campaigns, discussion of policy and function and so on. They may contain words, pictures, diagrams, cartoons etc. or any combination of these, depending on what is suitable for the audience and the message. There are many kinds of visual aid to choose from. We have arranged them in three groups: projected aids, three-dimensional aids and printed material.

Visual Aids: Projected

A projected aid is one that uses projection equipment such as a film projector, slide projector, overhead projector (OHP) or episcope. Large, bright images, particularly photographic ones, can be very attractive and powerful and filmshows often draw a large crowd, regardless of their subject matter. Most projected aids, however, are expensive and complex to produce and demand special viewing conditions.

Advantages:
— they are good for realistic illustration (e.g. photographic — shown in film, slides, or with an OHP or episcope);
— they can convey strong messages with high emotional appeal (mainly applies to film);
— they are good for showing examples and case studies;
— they can show movement (film; OHP for simple animation only).

Disadvantages:
— they are often expensive and complex to produce (film, some OHP and photographic slides);
— it may be difficult to pause, or to repeat pictures when needed (film, some slide projectors);
— correct viewing conditions and light levels may be essential for successful communication and difficult to achieve (film, photographic slides, episcope);
— projection equipment may be expensive or difficult to maintain;
— electricity is usually needed

Slide-tape

Projected slides can create an atmosphere, show familiar scenes, or show new places and new information through photographs, drawn pictures, diagrams etc. When combined with audio recordings, they can be a powerful method of communication. Members of the target group can be involved when photographs are taken locally, and they can take part in the making of a slide-tape programme if the producer draws on the talents of the community for acting, song, interviews and so on. Slide-tape can be arranged in different ways. A simple presentation might mean somebody operating a slide projector according to requests or cues on a sound tape. A complex presentation can involve a complicated arrangement of linked equipment. Its advantages and disadvantages combine those of projected aids and sound recordings, but here are some important ones:

Advantages:
— slide-tape can combine the strength and impact of photographic image with recorded sound;
— it can show things realistically, both familiar scenes and new information.

Disadvantages:
— it needs both projection and sound equipment;
— presentation can be difficult to arrange unless it is designed to be very simple.

In the picture, a slide sequence shows local people another part of their country; they get a vivid impression of an area where villagers have installed their own piped water supply, so that women no longer have to trudge to the next valley to collect water. The associated sound includes villagers talking and singing about their achievements, and a commentary.

Chapter 5 Narrowing the Field

Visual Aids: Three-Dimensional

The use of models, games and puppets has already been discussed on pages 42 to 44.

Visual Aids: Printed Material

This group includes many different aids. Each kind can be produced as a single piece of material for some specific use, but is usually more effective if duplicated or printed, so that copies can reach a wider audience. The advantages and disadvantages of printed material in general are given in Chapter 6. One important point is that paper can easily get crumpled, torn or marked unless it is looked after, stored carefully when not in use, or given a protective covering.

Poster

This is a large sheet of paper carrying a bold, clear message — one single idea is best, conveyed in a brief slogan with a maximum of seven words. It should have enough impact to catch the attention of a random passing audience. It aims to create awareness or promote action (it may persuade, warn, forbid etc.), and its meaning must be self-evident.

Advantages:
— it is good for giving people simple, straightforward information, telling them of a filmshow or other forthcoming event;
— it is good for reminding people of what has been taught, e.g. as part of a campaign.

Disadvantages:
— it is not suitable on its own for teaching and explaining;
— it is easy to forget that poster messages do not have lasting impact; once people get used to seeing a poster, they stop noticing its contents.

Chart

This contains much more information than a poster, and is normally used where there is a captive or motivated audience such as a class of learners, or patients waiting at a health centre. A chart may be designed as:

• a **teaching chart** to aid the instructor's presentation, bold enough for all learners to see.

Advantages:
— it is good for presenting an overview;
— it can show a step-by-step process, although the whole process will be visible from the start unless parts are covered up;

• a **wallchart**, with detailed information suitable for study by individuals or groups. It may be left up as reference for some time, for viewers to look at several times, carefully and at close range.

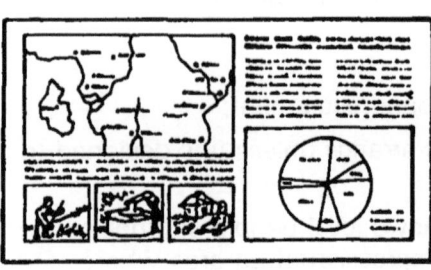

Advantages:
— it is good for showing background information not central to the instructor's theme;
— it is good for a long-term source of reference (e.g. for signs and symptoms of diseases, different kinds of implements and their uses).

Calendars are often used in a similar way to charts. In addition

Chapter 5 Narrowing the Field

to telling you what day it is, they can give other more important information, such as farming activities or facts about health.
They may be designed as a single sheet that lasts all year, or as a series of pages presenting new information each month.

Flip-chart

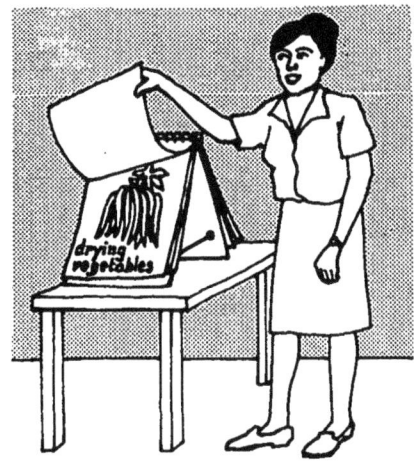

This is a series of pages, charts or large pictures, fastened together or held between stiff covers so that the material can be propped up or hung over a board for presentation by the instructor. Information is presented in sequence, page by page: the movement of turning the pages helps to draw the audience's attention to each new point.

Advantages:
— it is suitable for information which can be broken down into chunks, e.g. different aspects of a topic or individual steps in a process or system;
— it is good for learner participation, because the pace of learning can be varied, and pauses can be made at any stage for discussion or explanation.

Disadvantages:
— it can be difficult to handle if it is not well-designed and produced; (comb binding or spiral binding are best — see Book 2).

Flannelgraph

Materials with rough or hairy surfaces tend to stick to each other. A flannelgraph consists of cut-out words or pictures backed with sandpaper or a hairy or rough-textured cloth; when pressed on to a sloping board covered in material such as flannel, blanket, felt or winceyette, they will stay in place on the background.
A similar idea is used for other kinds of display board (magnet board etc.), but a flannelgraph is the cheapest and most adaptable.

Advantages:
— it is good for building up information and developing a theme; it is a very flexible medium — pieces can be placed on the board when and where you want them, and removed as desired;

— it is a useful medium in remote areas, as it is easy to improvise — a dark blanket hung over a board makes a good background; coarse sand or soil can be glued to the back of the cut-out pieces;
— it is good for lively audience participation, because the pace of learning can be varied and pauses made at any stage, and because learners can easily use the flannelgraph themselves.

Disadvantages:
— since the pieces are light, they can get blown off the board by draughts from ceiling fans or windy weather;
— the pieces can be lost or damaged — they need to be stored carefully, and preferably kept in the right order so that they are ready to use.

The picture shows cut out pictures and arrows used to illustrate the movement of produce from farmer to consumers.

Word or Picture Cards

Pieces of card with words or pictures may be used in a number of ways to stimulate learner activity, particularly in literacy or language work:

Chapter 5 Narrowing the Field

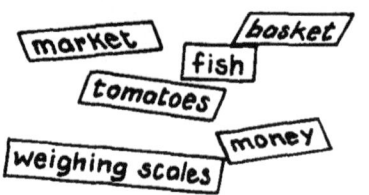

- question and answer
- word recognition, word cue
- sentence formation
- vocabulary

Picture cards have other uses:
- to promote discussion, provoke thought
- as a basis for writing or further activities
- to help show or explain something
- to help illustrate a difficult concept.

Cards held up quickly for brief illustration or immediate response are known as 'flash cards'.

Hand-out Sheets, Newsletters

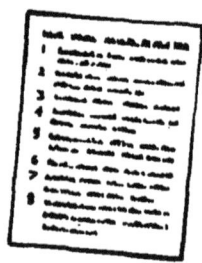

These are sheets of paper on which information is printed or duplicated. They are usually cheap to produce and can be distributed in large numbers.

One picture shows a local newsletter distributed to co-op members to give items of current news and general interest. The second example is a simple hand-out sheet summing up the responsibilities of a co-op member.

Leaflets

Leaflets are similar to hand-out sheets and newsletters, but are usually folded. They can only present limited information — overcrowded contents will look confusing and will not be easily understood. They are useful as reminders, e.g. handed round after a talk or given out during a campaign.

Booklets

A booklet consists of up to about twenty-four pages joined together at the spine; it may be small in size and have few pages, but it has more space for information than a leaflet, and can examine a topic in greater depth and detail.

Advantages:
- booklets can be used for teaching and learning sessions, for individual study, or as an instruction manual or practical workbook;
- they make a good and fairly permanent record of information, e.g. for reference or entertainment.

Pamphlets are a kind of booklet, but they are not joined together at the spine. Comics are also similar to booklets.

Case Study Exercise

Now let us consider the different methods and media in relation to a set of imaginary circumstances. A simple role-playing exercise gives the opportunity to apply some of the things we have discussed to a communication task.

In this case study the inhabitants of Tindul village face a number of problems. Paul, the district health worker, has made several visits to the village and spent time there in order to get to know people and find out what their needs and priorities are. Having done his research, he is now planning an educational programme

Chapter 5 Narrowing the Field

for Tindul. Which of the methods and media discussed will be most suitable for his work in the village?

Methods and media are listed in the table on page 55, with some of the advantages or disadvantages that would apply in these circumstances. Readers may wish to work through the table for themselves; first study carefully Paul's summary of the information he gained about his audience and the messages he decided were most appropriate.

Paul's Summary

Target Audience

Location and resources

— Tindul is remote and inaccessible; people travel on foot or sometimes by bicycle. Paul himself has a bicycle. The village can be reached by truck during the dry season, with some difficulty, but the track is poor and quickly gets waterlogged when the rains come.

— The main contact with the surrounding area is through the market held at the district town one hour's walk away. This is held every week, but villagers cannot attend regularly.

— There is no electricity.

— There is a village hall where meetings, parties etc. take place.

— Trees grow locally, so there is a source of building material and firewood; things which cannot be grown and made in the village must be brought from the district town.

— Soil is reasonably fertile in the area; food crops are grown, and people herd cattle and goats. Local diet is adequate. People can sell produce when they manage to reach the market.

— A nearby stream provides a source of water, but is not suitable for boats.

Education, Media and Ideas

— There is a high rate of illiteracy (62 per cent in the area), but a few adults can read, and those children who attend primary school at the next village.

— People rarely see books, pictures or newspapers, though a copy of the provincial newspaper is sometimes brought back from the town after market.

Chapter 5 Narrowing the Field

— The village headman has a transistor radio; this is out of use for long intervals when new batteries are needed.
— There is a tradition of story-telling and song.
— The headman has an influential position. He is alert for the welfare of his villagers, and not opposed to change if he considers it will bring benefits. He gets on well with Paul, and supports the ideas that Paul is hoping to introduce at the present time.

Subject Matter

The most serious problem in Tindul is that of ill health; villagers are upset by the amount of sickliness and early death among their children. Paul has traced many of the diseases and parasites to poor hygiene and sanitation, and decides to concentrate on ways of preventing illness through improved practice in this area. His main messages will be:

— keep food covered
— prepare food on a raised stove
— boil drinking water
— wash your hands properly and at the right times
— keep all your body clean
— dig a rubbish pit and cover the rubbish with a layer of soil each day
— build latrines, use them properly and keep them clean.

Chapter 5 Narrowing the Field

Tindul: Methods and media for an educational programme

Method/medium	Advantages	Disadvantages
1 Learning through achievement		The village is too inaccessible for Paul to keep returning for this kind of constant instruction.
2 Models		
3 Games	If Paul could think of a suitable game this would be a good way of reinforcing messages. He is looking out for games played in the village; and asking friends and colleagues for ideas too.	
4 Drama	This is too ambitious for Paul at the moment, but if messages in stories and songs are effective he will think about trying some acting games and drama.	
5 Puppetry		
6 Story-telling	A local tradition already exists, so this would be a useful way of communicating messages, and would use existing communication networks.	
7 Song		
8 Discussion	An essential part of Paul's educational work	
9 Field trip		Inaccessibility makes group visits outside the village impractical.
10 Demonstration	Demonstrations are appropriate for some of the messages.	
11 Talk or lecture		
12 Exhibition		
13 Sound recordings	The villagers sing well and could compose songs specially to convey Paul's messages. Perhaps he could arrange to exchange their songs and messages with other villages.	The headman's radio is unreliable, because of the irregular supply of batteries. Paul could bring a radio or tape recorder with him, but these are rather heavy and inconvenient to carry.
14 Visual aids: projected		No electricity.
15 Slide-tape		
16 Visual aids: printed material	Pictures would help Paul convey some of his messages, e.g. flip-chart, flannelgraph, picture cards. Leaflets and posters could be left behind at Tindul, to reinforce messages. Paul is thinking of encouraging villagers to produce materials themselves; he is finding out about low-cost printing methods.	Low level of literacy and unfamiliarity with pictures must be taken into account. Charts might present too many complex relationships between words and pictures. Some materials will be heavy or inconvenient for Paul to carry.

Chapter 5 Narrowing the Field

Section C

Using Printed Material

Chapter 6   Why Use Printed Material?

What is Printing?

One of the oldest and best established of media, printed material is the most widely known, widely used and generally accepted. Different printing processes work in different ways and it is difficult to give a definition which is generally applicable. However, printing can normally be defined as 'the production of multiple copies of an original image, usually using ink pressed on to paper' For the purposes of this package, we are interested in the techniques most appropriate for the communication work of field officers, extension workers, adult educators, teachers and trainers. These are likely to be low-cost methods of producing materials such as leaflets and booklets.

Advantages of Printed Material as a Medium

- There is a wide variety of different printing processes, allowing the choice of the most appropriate for any given situation. This variety makes possible other advantages:
- A wide range of possible formats (hand-outs, workbooks, posters etc.);
- Printing on to different surfaces and materials (paper, cloth, plastic packaging etc.);
- Printed material can be used in a number of different ways. It can be used as a medium in its own right, or as support for other kinds of media, e.g. in campaigns. It can be something with a short life, read and thrown away, or it can be a permanent record to be used and re-used. It can be designed for individuals (e.g. a pocket book) or for groups or crowds (e.g. a poster pinned up for all to see).
- Printed material can be produced to any level of sophistication and finish and suited to any audience.
- Production facilities needed will vary according to the process used and the standard of results required. Material can be produced without access to electricity.
- Printed material can be used and viewed freely; no special rooms or facilities are needed (sometimes not even a classroom).
- Readers can use printed material in their own way and at their own speed — they can learn at their own pace and read things over again.

Disadvantages and Problems

- Printed material may be fragile and susceptible to wear and tear.
- It can be difficult to store as it may come in many different shapes and sizes.
- Long-term storage may be especially difficult due to bulk and susceptiblity to damp, heat, dust etc.
- Print can sometimes be costly, e.g. if elaborate materials and special finishes are required, or if paper is in short supply.
- Distribution may be difficult.

Chapter 6 Why Use Printed Material?

- With a largely non-literate audience, special care must be taken to create material which is meaningful and interesting.
- Printed material may be rather impersonal if used on its own; it is easier to identify with a face-to-face approach or a voice (e.g. on radio) than with the written word. Print, like other media, can probably be most effective as part of a multi-method/media approach.

Low-Cost Printing

The effective use of low-cost printing can play a crucial part in both rural and urban development. By 'low-cost' we mean inexpensive in capital outlay and running costs, and using skills which can be found, or easily taught, in places without access to sophisticated technical resources. The printing techniques used in such operations include hectography, ink and spirit duplication, screen printing (also called silk-screen printing), hand-operated letterpress and simple offset litho printing. They require basic training, practice, and a careful, common-sense approach, but not a long and expensive technical training. Much of the material produced is likely to have a short life: posters, leaflets, workbooks, booklets, rural newspapers and so on.

Methods which are low-cost in one place, of course, may be costly in another. Sometimes less commonly used methods of printing might be appropriate, for example in situations where it is difficult to obtain the usual printing machinery and materials. These might include, for instance, the use of locally produced dyes and woodblocks for printing on to cloth. Whatever the process, printing equipment and other resources must be used efficiently and to maximum effect in order to be run at low cost.

Printed material will normally play a part in any project or campaign. Those who are closely involved in such work can often sense the growth of local interest in a particular topic, or foresee the likelihood of such an interest, which may create a need for printed material. For instance, if farmers are considering whether or not to adopt a kind of seed or a practice which the Ministry is recommending, if women are curious about immunization or family planning, or if school-leavers wish to know about training or bank loans, then the need for some literature may be evident.

Examples of materials which are often needed include: literacy continuation materials; leaflets giving information on nutrition or child care, or on how to apply for agricultural grants; notices giving warning of aerial crop spraying; children's storybooks and workbooks; machine operation manuals. Many of these need to be produced locally, in small numbers, so that the writing and design can take into account factors such as language and agricultural conditions, which vary from district to district.

Centralized and Local Production

Within any organization, system or Ministry, printed material is usually produced centrally at national headquarters, state level resource centres or government printers; less frequently is there any provision for the production of materials on a local basis. Centrally produced material can be designed and printed to a high standard, and presented in a sophisticated and attractive manner, calling for a higher level of staff ability, resources and funds. It can provide authoritative materials and accurate, up-to-date information in line with current policy.

We have, however, already come across reasons why printed material might be needed at grass-roots level, and why local factors might make local production desirable. Locally produced material can provide elements that centrally produced material cannot, and can therefore contribute usefully to the work of communication. For reasons of time and economy, materials produced centrally must be printed in large numbers and would therefore normally be aimed at a nation-wide audience; they cannot afford to take into account the precise needs and characteristics of each small target group. Locally produced materials could support and extend the effectiveness of centrally produced materials in the following ways:

— they can be produced in small quantities to meet a specific local need;

— they can be revised relatively quickly in reponse to the local situation, bringing immediacy and impact;

— they do not have to be produced to the same level of sophistication and finish as centrally produced materials, and so will be less complicated and costly to produce;

— they can be written in the local language or dialect;

— they can be pretested on the spot and discussed with the people who will use them;

— local people can be involved in the design and production of materials, thus ensuring interest and a certain amount of commitment;

— by including accurate local detail in text and illustrations, the target audience will more readily see the relevance of the material to themselves.

Since local producers are closer to the local audience, they have more opportunity to research their target groups, find out needs and priorities, and make materials relevant, interesting and effective for local people to use.

Some Alternatives to Using Printed Material

Check these before deciding to produce printed material.

• Can you improve face-to-face communication by, for example:

— home visits and talking to people (however, printed material such as picture cards may be very useful to take along on home visits);

— calling a meeting to discuss the issue (printed material such as hand-out sheets may again provide useful back-up, and help to spread the information beyond the immediate participants).

• A different medium might be more appropriate (e.g. can a picture show the subject clearly enough or would a three-dimensional model be better?) used instead of, or together with, printed material.

• Have you considered using a wall newspaper? If there is a central site where written messages can be pasted, pinned up or even painted on a wall, there may be no need to distribute multiple copies. Single copies of written news items, information sheets, explanatory pictures etc. could be put up, for example on the wall of a village meeting place, for people to read each evening as they return home from the fields.

Chapter 7

Forming the Message

Whatever medium is used, the important thing is that it should help to communicate the message effectively. The material must:

- attract the attention of the audience
- hold it long enough for the audience to take in information
- convey information clearly so that it is easily understood.

We have taken printed material as our medium; we must now explore how best to use words and illustrations in order to make the message clearly understood and favourably received. It is important to match the content and style of your material to:

- the audience
- the printing process (when you do not have a choice of process).

What you need to know about an intended audience is discussed in Chapter 2 of this book; different printing processes are described in Book 3. This chapter looks at content and style of material, at written text, and illustrations. It aims to give a few useful guidelines for good practice, to point out potential problems, and to show how material can be checked at the draft stage to make sure it is appropriate.

Contents

Information

Pay attention to the way in which information is presented to the audience — the amount, the order and the way it is laid out on the page. Beware of overloading readers with too much information, too densely packed.

Density

How much is new to your readers? Try to lead them on a guided journey: do not overwhelm them with solid pages of indigestible facts. Closely written, unbroken text looks difficult to read and can be very off-putting. Split new information into manageable pieces; break it down into sections, paragraphs, lists etc; space it out with explanation, headings, summaries, relevant pictures and white space.

Include all the essential facts, but do not be afraid to leave out other information which may distract the reader from what is really important. (It is not always necessary to write down everything you know about a subject!)

Sequence

Think about the order in which facts are presented. It is important that readers know where the information is leading, what it relates to and what they are to do with it. If there is a 'right' order (e.g. to explain the safe use of a machine, a specific movement, a procedure that must be followed), give information in that order. Writing instructions in an inappropriate order will certainly be confusing — it may even be dangerous! Compare the two sets of instructions for making fried maize meal bread.

Chapter 7 Forming the Message

To make fried maize meal bread

1 ½ tsp salt
3 tsp baking powder
2 cups maize meal
1 ½ cups milk

2 eggs
2 tsp fat
2 tsp sugar

Mix together the dry ingredients.
Add the milk, eggs (beaten) and fat (melted), and mix well.
Pour batter on to a hot, greased frying pan or griddle, making cakes about 3 inches across.
The pan is hot enough when a drop of water dances about on the surface.

To make fried maize meal bread

1 ½ tsp salt
3 tsp baking powder
2 cups maize meal
2 tsp sugar

1 ½ cups milk
2 eggs, beaten
2 tsp fat, melted

Mix together the dry ingredients.
Add the milk, eggs and fat, and mix well.
Grease a frying pan or griddle, and heat until a drop of water dances about on the surface.
Pour the batter on to the hot surface, making cakes about 3 inches across.

Facts are best stated in one of three ways: chronological, known leading to unknown, or simple leading to complex, as shown in the following examples.

- Chronological — follow the order in which things happen, or in which things should be done, from beginning to end.

Rice Harvesting

Chapter 7 Forming the Message

- Known to unknown — start with information already familiar to readers, where they feel confident, and from there lead on to explore new ideas.

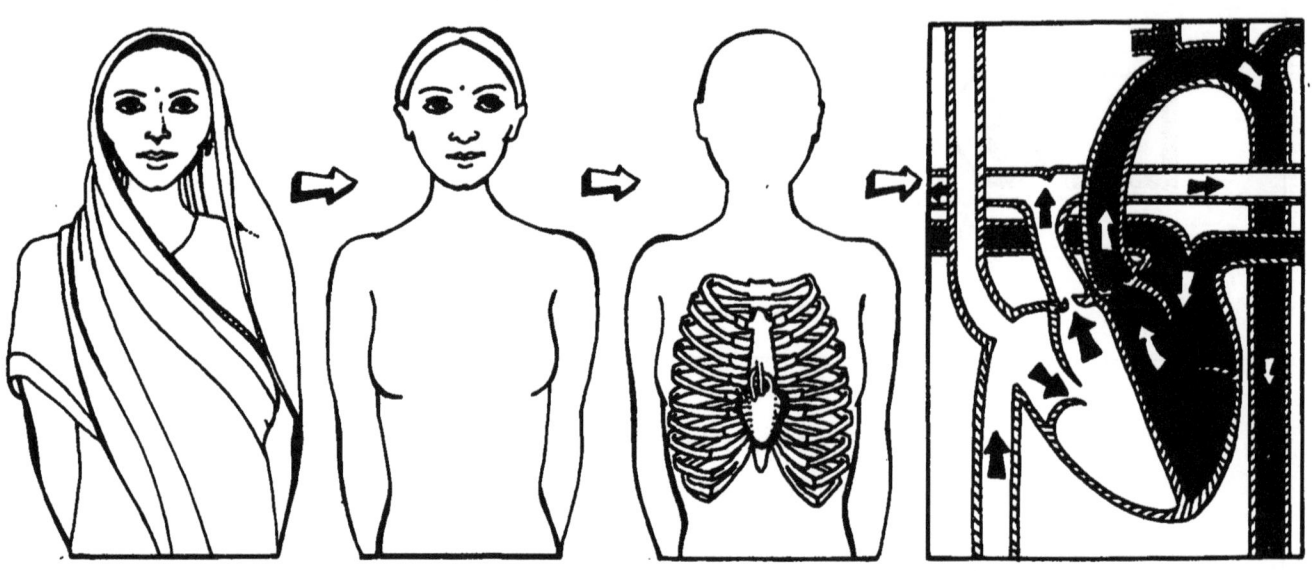

- Simple to complex — it is better to lead up to something difficult by degrees, perhaps explaining a piece at a time. Do not begin with the most difficult part, or with a really complex description of a whole system.

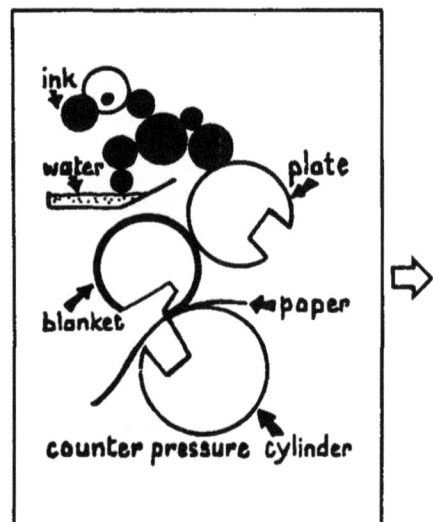
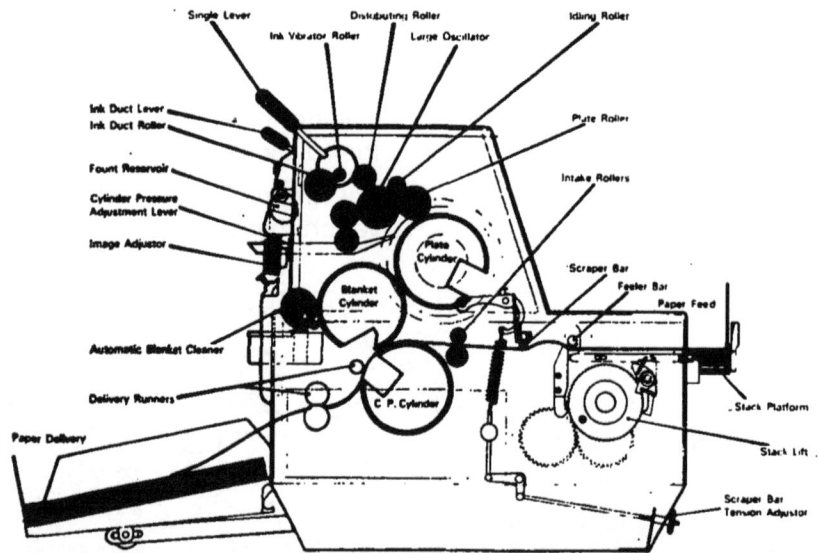

Thinking of a good sequence will help to break the information into manageable chunks. 'Signposts' can be used — visual cues which alert the reader to new information or a new activity. For example, headings indicate a change of subject or emphasis, a change of layout or lettering can warn readers of questions to be answered, a picture of a pair of scissors might draw attention to where a paper should be cut.

Approach

Think about suiting the contents to the readers in terms of language used. Remember which language or dialect they are familiar with and can understand, but also think about the writing style, vocabulary and general approach.

Chapter 7 Forming the Message

An official directive may be what you are used to receiving yourself, but it will not always be the best way of informing and instructing other people. It may be necessary to persuade them, giving reasons and explanations — or even to inform indirectly by entertaining or amusing them.

Style

On what level do you want to appeal to your readers?
Your approach might be:

— formal (a correct, official style, seeking to impress or gain approval, as in dealings with local government, applications to Ministries, or addressing respected community leaders)

— informal (relaxed, direct, not standing on ceremony)

— colloquial (conversational, chatty, probably using local dialect, idiom or slang phrases)

— personal (intimate, friendly, everyday talk as with family and friends).

For those who write in a formal style for their official business, it may be very hard to write in any other way. It is easier to talk than to write informally, so it may help to tell someone your message, and then write it down the way you said it.

It is essential that the material can be easily understood — otherwise there is no point in writing it. Unless there is some special reason to follow a traditional (e.g. ballad or story) style or to use ambiguous language, material should be written simply in a straightforward and uncomplicated way. It is helpful to:

— be absolutely clear about what you want to say and why, before beginning to write;

— say it clearly and directly; it may help to write something down roughly, and improve on it later; read it through aloud to see if it flows naturally;

— test what you have written — get others to read it and check that they can understand what you mean.

Writing Simply

Here are some sentences taken from draft material intended for literacy learners. They can all be written more simply so that they are easier to understand.

1 Trees have been identified as being indispensable to people.

2 Every location is supposed to have a tree nursery.

3 The efforts of those who helped to initiate this book cannot go unthanked.

4 One spraying is done immediately after the first weeding and the second is executed after the second weeding.

5 Grade one potatoes are put in sacks and then transported by any available means to a market for sale.

6 It should be remembered that the local people who depend wholly on the coconut tree will never at all draw their firewood from the trunk of this tree.

7 Weeding starts after the plants have grown to some reasonable height (15 cm).

Chapter 7 Forming the Message

These sentences can provide some examples of points to keep in mind, particularly when writing for newly literate and semi-literate readers.

(**Note:** the following guidelines refer to writing in English. For materials written in other languages, check that each point is appropriate and make changes as you think fit.)

General

- Material is readable if it can be read and understood by those for whom it is intended. Thus, if the teacher or extension worker has to *explain* the language, it is not well written for the purpose, even if the audience understand the explanation.

- You must decide whether the reader is going to try to read the material unaided, or whether someone will be there to help.

- Do not feel that the message always has to be obvious. If you can apply the phrase 'of course everyone knows that . to what you are saying, maybe there is no need to say it at all.

Choice of Words

- Use concrete words — words with an immediate meaning: 'handspan' (illustrated) might be better than '15 cm'; 'reasonable height' is too vague (sentence 7).

- Use short, simple words — 'people need' rather than 'indispensable to people' (sentence 1).

- Concepts — 'trees have been identified'; avoid such general notions (sentence 1).

- Familiarity — use common and familiar words wherever possible. 'Write and make this book' is preferable to 'initiate . . . (sentence 3). If a specialist or technical term or unusual word is really needed, make sure that it is acceptable and then use it consistently. Explain a new term fully with relevant examples and pictures.

- Vocabulary — do not be afraid of using a simple word more than once: there is no point in varying them unnecessarily. In sentence 4, 'done' could be repeated, rather than introducing 'executed' later in the same sentence.

- Local vocabulary — look out for words related to local culture, or to topics which are important in the district. For example, there may be special terms for herds or crops.

- Colloquial, local words — popular slang and idiom can be very effective, and make material more interesting for local readers. Do not feel you must protect your readers from 'incorrect' grammar and use of language; but be sure that slang expressions are appropriate for the intended audience, and remember that people from other groups may find them difficult to understand.

Sentences

- Keep sentences short.

- Keep sentences simple. Avoid constructions such as: 'It should be remembered that the local people who depend wholly on . . . will . . .' (sentence 6).

- Use active verbs. 'Ask the Agricultural Extension Officer to help you plant a tree nursery' is better than 'Every location is supposed to have a tree nursery' (sentence 2). 'Take your potatoes to market' is better than 'Potatoes are . . . transported by any available means to a market' (sentence 5).

- Positive statements are better than negative ones. Use 'We want to thank . . .' rather than 'Those who helped . . . cannot go unthanked' (sentence 3).

- Personalize the writing: 'Start weeding . . .' or 'You can start weeding . . .' is better than 'Weeding starts . . .' (sentence 7).

Avoid very long passages of continuous prose; they can be tiring and boring to read. If a paragraph takes more than half a page of your booklet, it is probably too long. You may prefer not to use formal sentences and paragraphs all the time; there are alternatives, including:

— lists

— questions and answers

— examples and explanations set apart from the text

— tables (e.g. for comparison)

— charts (e.g. for choosing from alternatives, following steps in a sequence)

— introducing activities, demanding a response from readers

— using visual effects (e.g. emphasizing key information by putting it in a box)

— using appropriate pictures and diagrams which will bring variety to your material and help to make it clear and interesting.

More information about the presentation and arrangement of written information is given in Book 2.

Choosing Illustrations

Picture Recognition

Everyone finds pictures appealing — even someone who cannot read. However, if people seldom see pictures and are not used to relating two-dimensional images to real life, they may find a picture difficult to recognize and its message difficult to understand. Here are some guidelines for making pictures as easy to understand as possible.

1 Content:
what the picture is about may be even more important than the way in which it is presented. A subject relevant to the audience is more likely to be understood than one which is of only casual interest.

2 Familiarity:
people are more likely to recognize realistic pictures of things they know well, and will identify more easily with pictures showing local life and a familiar setting.

3 Human activity:
pictures are most often successfully understood if they show somebody doing something. People are interested in other people; they normally look at a person in a picture first, to see what he or she is doing.

Chapter 7 Forming the Message

4 Whole objects:
pictures of whole objects will usually be easier to recognize than pictures showing parts of things; for example a hand or head alone may not be clearly understood.

5 Detail:
take care with the amount and accuracy of details you include.

• Although details can help to make a picture realistic and interesting, too many unnecessary details can distract from the point of the message and confuse the whole issue. Of pictures (a) and (b), people found picture (b) more confusing.[1]

(a) (b)

Only include what is essential for communicating the message and keep backgrounds as simple as possible. (The amount of essential detail will be influenced by the intended audience, and by the way in which the picture is to be used — a teaching aid or discussion picture will probably include more information than pictures designed for learners on their own.)

• Accurate, realistic detail can assist the recognition and understanding of pictures; inaccurate detail may distract or confuse so that the intended message is misunderstood. This picture was made in Keyna.[2] The fact that the tail turned downwards made the drawing a *cow*, not a *goat*, because goats' tails turn upwards.

6 Perspective:
drawings using perspective, or showing objects from unfamiliar angles, may be difficult to interpret.

Drawings[3] used in tests:

in Kenya   in South Africa

 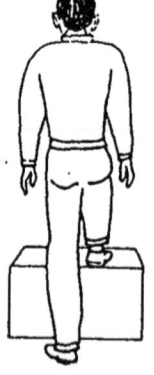

'Which is nearer to the man, elephant or antelope?' Both answers were given.

'What do you see in this picture?' Some saw a man climbing a step, some saw a man with a maimed leg.

Chapter 7 Forming the Message

'We don't have mosquitos that big in our village.'

Hand wash   Hot iron

Can be dry cleaned   Can be bleached

7 Size and scale:
objects shown much larger than their usual size can be difficult to recognize, or the picture may be considered irrelevant.

8 Colour:
should be absolutely accurate, if it is used. It can sometimes clarify pictures, but it can just as easily distract and confuse. People often enjoy bright colours and prefer coloured drawings, but using colour does not necessarily make recognition and understanding easier.

9 Signs and symbols:
must be taught and explained just like written letters. This applies equally to pictures used in a symbolic way, to abstract signs and symbols and to diagrams, graphs, maps and other symbolic ways of visualizing information. If symbols are not already familiar to the audience, they will be difficult to interpret. For example, do you understand these symbols? They are washing instructions shown on labels of clothes sold in the U.K. The first two symbols are at least based on recognizable objects; the last two are abstract and therefore cannot be understood at all without some explanation. Guidelines for using symbols follow on page 71.

10 Sequences of pictures:
relating a picture to other pictures before and after it is also an acquired skill. Someone unfamiliar with the concept will not be able to understand the linking of two different ideas in a single picture, or the jump from one picture to another in a picture story or cartoon strip.

Drawings used in a study in Nepal[3]:

Almost half the respondents thought that the pictures were of different women. Told that it was the same woman, most people could piece together a sequence of events.

Picture Style

Illustrations can be presented in a number of different styles; the choice of style influences people's preference for one picture over another, but it can also affect recognition and understanding. Several studies have been made about the effect of art style on preference, recognition and understanding, but the results have not been conclusive. The most appropriate style varies according to the target audience and the kind of picture, as well as depending on the resources and printing process available, and the quality of image that can be produced. Here are some different styles: stick figures, silhouette, simple outline, line drawing, photograph and 'blocked-out' photograph (photograph with the background removed).

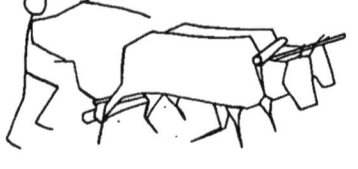

Stick Figures
— the most simple shorthand style, though not always as easy to use as it may seem;

Recognition
— they are fairly easy to recognize once the style is familiar, but care is needed in case the angle of view or background information is confusing;

Popularity
— they are not generally popular with audiences, because they have no emotional appeal and are difficult for audiences to relate to;

Production
— they are easy to draw and easy to reproduce.

Silhouette
— a bold, simple, solid shape; strong contrast gives it impact;

Recognition
— very simple shapes are fairly easy to recognize; it is not suitable for unusual angles of view or complex overlapping shapes;

Popularity
— it is not generally popular with audiences, though in some places it may relate to traditional media such as shadow puppets;

Production
— the outline needs careful drawing to explain what is happening, since all detail inside the shape is lost; it is easy to reproduce.

Simple Outline
— a much simplified version of a realistic image;

Recognition
— it is fairly easy to recognize;

Popularity
— it is not generally popular with audiences;

Production
— it is fairly easy to draw and easy to reproduce.

Line Drawing
— a simpler and bolder version of an accurate, realistic image;

Recognition
— it is generally easy to recognize — easier than a simple outline, maybe because it contains more detail;

Popularity
— it is generally quite popular with audiences;

Production
— some artistic skill is needed; it is easy to reproduce unless it contains very fine detail.

Chapter 7 Forming the Message

Realistic Drawing
— an accurate and realistic style, using tones and shading;

Recognition
— it is easy to recognize;

Popularity
— it is generally quite popular with audiences;

Production
— artistic skill is needed and it is complex to reproduce.

Photograph
— accurate and realistic;

Recognition
— it is generally easy to recognize, although the background may be confusing and there may be unnecessary details such as trees and shadows;

Popularity
— it is generally popular with audiences;

Production
— it is expensive to produce and relatively complex to reproduce when used in printed material.

Blocked-out Photograph
— accurate and realistic;

Recognition
— it is easy to recognize — easier than ordinary photographs, because confusing background is removed to clarify the main subject matter. However, this can sometimes destroy valuable information, and can leave confusing shadows etc. with no apparent reason or meaning;

Popularity
— it is generally popular with audiences, tending to be more popular than drawings;

Production
— it is expensive to produce and relatively complex to reproduce when used in printed material.

From these brief notes we can see that there is no single best style, and that all styles have some disadvantages. Generally speaking, the most useful, easily produced and acceptable are simple outlines and line drawings. For further information see Book 2, Chapter 3.

Symbols

Symbols are relatively simple things which are used to represent something more complicated, more abstract, or more difficult to understand — and certainly more difficult to illustrate.
Any picture involves some use of symbols, just by showing real life in a simplified, two-dimensional way.

Well-chosen pictures illustrate easily and quickly — often more quickly than words. Some symbols are based on realistic pictures and attempt to use the ease and speed of a picture, while leaving out the detail. For example, a stick figure is generally recognized as a man.

Chapter 7 Forming the Message

Local custom or convention usually determines the use of symbols. These examples are recognized in some places but unknown in others.

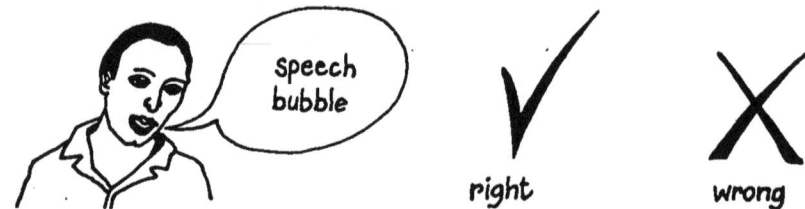

When ticks and crosses were not understood by village women in Uttar Pradesh, India, other symbols had to be found for 'right' and 'wrong'. Two birds were already regarded as symbolic by local people: the parrot was considered to be lucky, and the owl unlucky. The pictures[4] show how the artist used the parrot to mean 'good', 'do this' (feed your child while the food is still hot) and the owl to mean 'bad', 'don't do this' (do not give him food that has been stored). These symbols could not, however, be used in the U.K., where the owl is considered to be a wise bird and the parrot a foolish one!

Pictures which show concrete objects or combinations of objects can be used to symbolize other things, provided that their meaning is explained and is accepted. For example, a picture of a market could be used to illustrate a number of different subjects: villagers selling vegetables, a woman buying vegetables to provide a balanced diet for her family, members of a co-operative selling surplus produce to help raise money for a new plough. These are fairly straightforward. The picture could also be used, however, to represent something much wider and more abstract, such as 'commerce' or 'profit' or 'the rewards of hard work'.

Chapter 7 Forming the Message

Try to avoid ambiguity of meaning when using pictures in this way. To the artist, his picture of a man cutting down a tree may show the clearing of ground to build and cultivate, and thus symbolize 'national development'. Other people might not see this connection unless it was explained, and in an area without trees the symbolism would be meaningless.

Symbols can also be non-representational. We have already mentioned ticks and crosses, and some other examples are shown here.

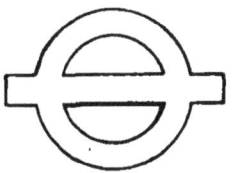 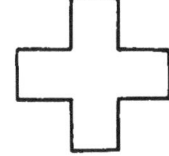 

London underground   *Red Cross*   *Red Crescent*

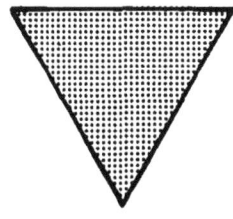

This triangle, coloured red, is seen on family welfare clinics in India. Although an abstract shape, with no obvious meaning, it has become familiar by repetition, and is accepted and recognized. The same triangle could, somewhere else, have quite a different meaning.

If you are using symbols, remember it is essential that your audience understand and accept your meaning. Where you can, use an existing convention. Otherwise, be careful to establish the symbol by explanation and discussion, or by associating it frequently with what it represents — perhaps over a long period of time.

Diagrams

Diagrams can be very useful for visualizing information which may be difficult or lengthy to describe in any other way, and can help people to remember it. Great care, however, is needed in their design and use. Most people will not be familiar with the symbolism involved — using diagrams is a skill that has to be learnt, just like using the written word. The easiest diagrams to explain will be those that use pictures or that contrast different sizes and quantities. Always keep the target audience in mind, and possible problems of recognition and understanding.

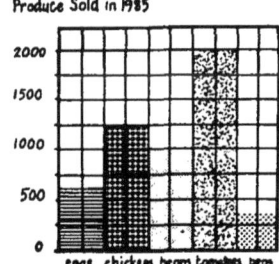

There are many different kinds of diagram. Here are some of the most commonly used: bar graph, pie graph, pictorial graph, flow chart, organization chart, sectional diagram, map or plan.

Bar Graph

— shows clearly the rise and fall of values or quantities of the subject.

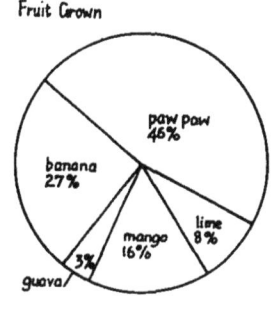

Divided Circle (Pie or Cake) Graph

— is divided into sections as in cutting up a pie. Each section is accurately drawn to represent a fraction or percentage of a whole. It can be clearly seen how these values relate to each other.

Chapter 7 Forming the Message

Pictorial Graph

— is adapted from the bar graph, but aims to be more interesting and attractive by actually representing the subject. A picture or symbol is used to represent a certain number of people or objects.

Families in the village

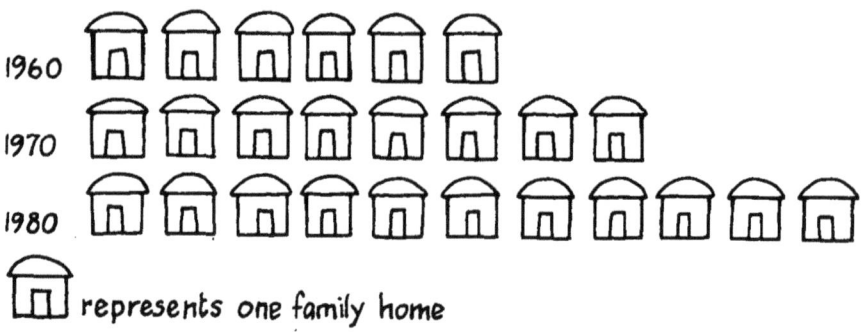

represents one family home

Flow Chart

— shows successive movements through a process from start to finish.

To Make Water Safe to Drink

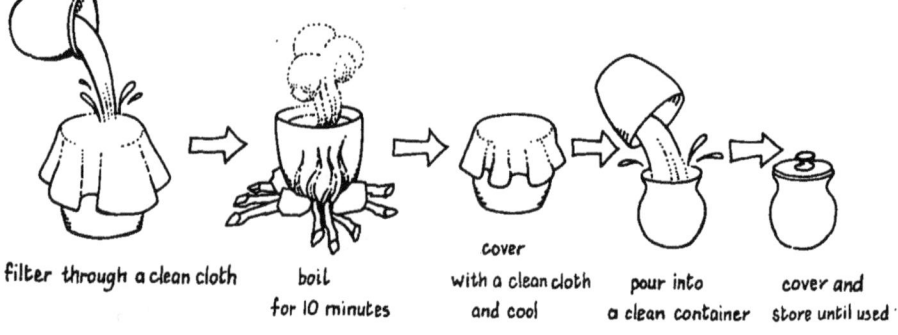

filter through a clean cloth → boil for 10 minutes → cover with a clean cloth and cool → pour into a clean container → cover and store until used

Organization Chart

— shows the interrelation of various units of an organization.

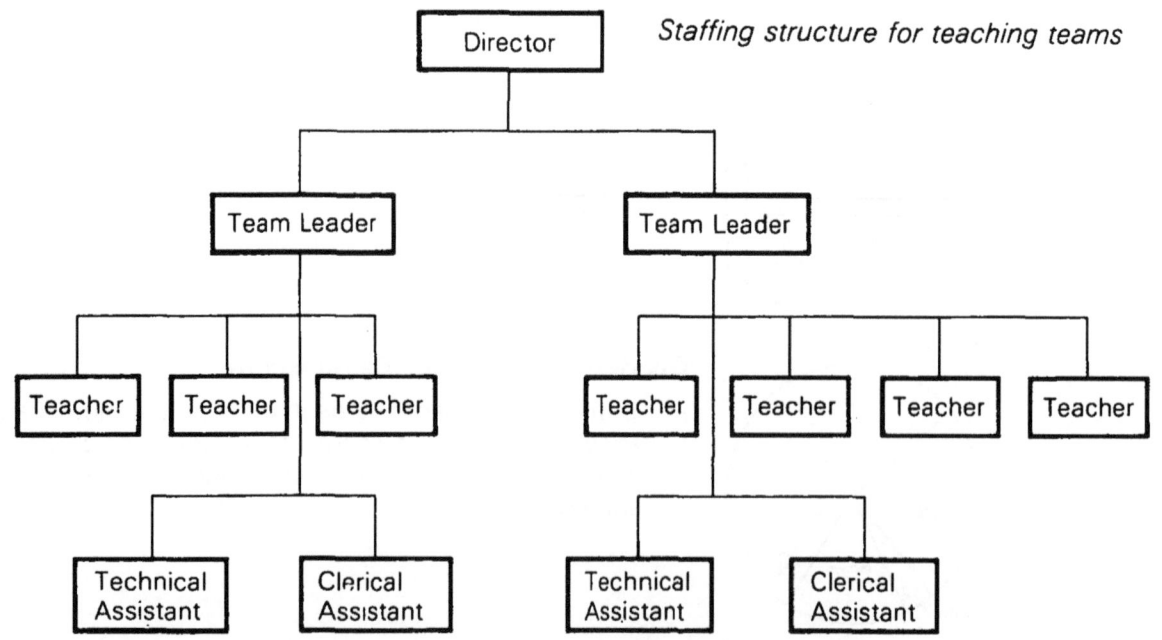

Staffing structure for teaching teams

Chapter 7 Forming the Message

A Pottery Cooking Stove

Sectional Diagram

— shows a simplified view of the inside of an object, to explain what cannot normally be seen. They are often used to clarify technical processes, the way things work and the way things are built.

Map or Plan

— looks down from above, to show arrangements, layouts or geographical features.

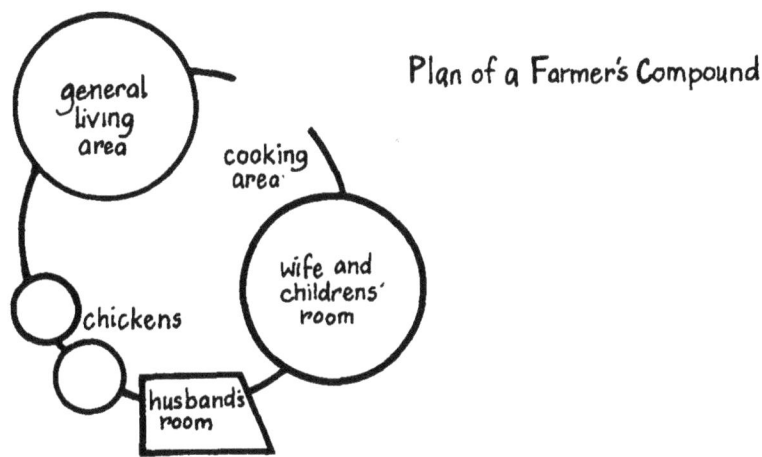

Plan of a Farmer's Compound

To avoid problems with understanding diagrams, test them before use, take time to explain them if necessary, and be prepared to stop using them if they are confusing. For more information about drawing diagrams see Book 2, Chapter 3.

Using Illustrations

Illustrations can make material look more interesting and attractive, and pretesting will help to make sure the audience can understand the pictures or diagrams to be used. Pictures scattered through a booklet for no purpose will just be distracting, however. Pictures use up time, space and money — they also need thought and planning. Remember that:

— illustrations should serve a purpose, such as showing something unfamiliar, stimulating questions; pictures which are too simple or obvious may waste valuable space;

— illustrations which are included in *teaching materials* can sometimes afford to be a little complex or ambiguous — for example, if they are intended to make learners think for themselves or work something out.

Pictures or diagrams can be used in the following ways:

- to stimulate interest or direct attention
- — attract interest
- — focus on an issue
- — stimulate discussion, questions or other learner activity
- to reinforce learning
- — test understanding (e.g. question and answer, identification of parts)

Chapter 7 Forming the Message

— repeat or summarize points made verbally (words and pictures can be used together for the repetition of important points)

- to help illustrate difficult concepts
— simplify and clarify something complex
— show something unfamiliar
— show something which cannot normally be seen
— explain something difficult to describe in words.

Let us look at some different ways of presenting information, using words, pictures, diagrams, tables and so on. Consider which examples work well and which are less successful; think how other methods or combinations of methods might present information more clearly.

Example 1: Tying an Arm Sling[5]

a) Numbered list of instructions

> **1** Support the forearm on the injured side, wrist and hand a little higher than the elbow.
>
> **2** Place an open triangular bandage between the chest and the forearm, with the point stretching beyond the elbow.
>
> **3** Carry the upper end over the shoulder on the uninjured side and round the neck to the front of the injured side.
>
> **4** Bring the lower end of the bandage up over the hand and forearm and tie off in front of the hollow above the collar-bone.
>
> **5** Bring the point forward and secure in the front of the bandage with a safety pin.

b) Pictures

Example 2: Best Months to Sow Vegetables[6]

a) List

> Cabbage — February to November
> Onions — February, March, April
> Peppers — January to June, and September to December
> Chillies — November, December
> Potatoes — May to November
> Beans — February to May, and October to December
> Spinach — March to May, and August to November
> Egg Plant — September to December

Chapter 7 Forming the Message

b) Table

	Jan	Feb	March	April	May	June	July	Aug	Sept	Oct	Nov	Dec
Cabbage		✓	✓	✓	✓	✓	✓	✓	✓	✓	✓	
Onions		✓	✓	✓								
Peppers	✓	✓	✓	✓	✓	✓			✓	✓	✓	✓
Chillies											✓	✓
Potatoes					✓	✓	✓	✓	✓	✓	✓	
Beans		✓	✓	✓	✓					✓	✓	✓
Spinach			✓	✓	✓			✓	✓	✓	✓	
Egg Plant									✓	✓	✓	✓

Example 3: Making Compost[6]

a) Prose with headings

> Compost is very important for the vegetable grower who has no other manures available. No vegetable matter should be thrown away but everything kept and put on the compost heap.
>
> *Site:* Choose a site which will not become waterlogged. This can be in the middle of the garden so that compost material does not have to be carried too far.
>
> *Size:* Drive four strong sticks in the ground, 120cm x 180cm apart.
>
> *Material to use:* Spread an even layer of grass, weeds, leaves or crop refuse about 25cm deep. Keep the edges straight and tidy.
>
> *The starter:* Spread a 10cm layer of kitchen waste or rotten fruit (e.g. mangoes), or animal manure or rich dark soil or old good compost. These contain bacteria which work on the material and turn it into compost. The 'starter' layer should be covered with another layer of material immediately, to prevent flies breeding in the starter.
>
> *Build up the heap:* Continue building up the heap in layers. Any ash available can also be added to the heap. It should be built up to 120cm high.

b) Pictures with labels

Base layer of refuse material: grass, weeds, leaves or crop refuse – spread evenly

Starter layer: kitchen waste, rotten fruit, animal manure or compost – cover immediately with another layer of refuse material

Continue building further layers of refuse material up to 120cm high

Chapter 7 Forming the Message

Example 4: Storing Onions[6]

a) Prose

> Store in thin layers in a well-ventilated room, in slatted boxes or on the floor. For onions which will be stored for some time, spray with malathion, against thrips. A good way of storing onions is to tie them on sticks and hang them up. This allows every single onion to be inspected. Look at all stored onions regularly and take out all those that are spoilt, otherwise you may lose them all.

b) Picture

Onions properly stored

Example 5: A Plastered Basket Tank for Storing Water[7]

a) Picture

b) Diagram with labels

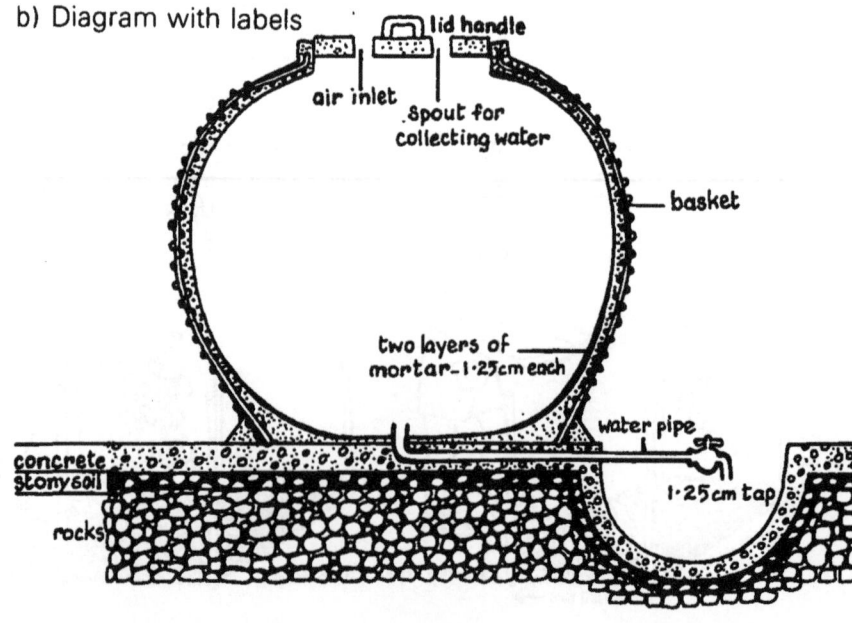

Readers may like to complete the table by adding their own comments.

Example	Version (a)	Version (b)	Comments
1 Tying an arm sling	Numbered list of instructions	Pictures	The pictures show most clearly what is to be done; the words do not add any useful information
2 Best months to sow vegetables	List	Table	
3 Making compost	Prose with headings	Pictures with labels	
4 Storing onions	Prose	Picture	Neither gives complete information, so both are needed
5 A plastered basket tank for storing water	Picture	Diagram with labels	

Words and pictures convey information in different ways; sometimes one is clearer, sometimes the other. In many cases, they can be used together for the most effective result. Common sense and information about the target audience must dictate when and how you use them, and what sort of style is appropriate. Pretesting will show if you are on the right lines.

Pretesting

Material should be pretested before it is produced and distributed on a large scale or at great expense. Pretesting measures the reaction of a sample audience group to the idea or piece of material being tested — it is really a trial run on a small scale. Pretesting does not guarantee that material will communicate effectively, but it does make it more likely.

It is difficult to test material objectively after you have put a lot of effort into producing it, and criticism will be hard to accept. For example, when people read the draft of this book and suggested changes, it was tempting to explain things away and discount the comments. When several people criticized the same things, however, it became clear that their suggestions deserved attention and that changes really were needed.

Chapter 7 Forming the Message

If it is possible, get someone who has not been closely involved in producing the material to do the pretesting — they will find it easier to get a true picture of people's opinions.

Pretests can do the following:

— show whether an idea is likely to be worth-while (e.g. testing the likely reception of material that is proposed);

— show whether partially completed material is on the right lines and is likely to be generally understood and accepted (e.g. testing a draft version);

— identify things that could be changed to make the material more effective;

— find out which of several alternative versions will be most effective.

Alternative Versions

Testing alternative versions is basically the same as testing a single version, except that the respondent must also be given the chance to choose one of several alternatives. The different versions may be arranged on a table, or in a rack, or pinned on a wall. Change the way they are arranged half-way through the series of interviews, to make sure that the arrangement is not influencing people's opinions of the material. Mark each version clearly with a letter or number and make a note of the arrangement people are seeing. The example shows four booklets arranged and rearranged on a table:

Arrangement for interviews 1-15

Arrangement for interviews 16-30

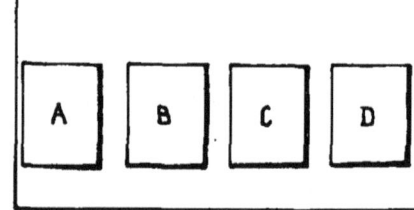
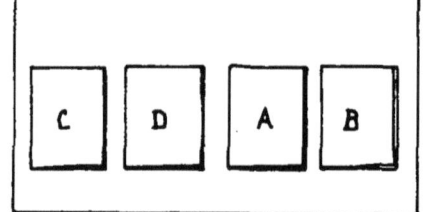

You can ask questions such as:

Here are four designs we are considering for a booklet cover.

Which do you like the most?

Why?

Which do you like the least?

Why?

We haven't decided which colour paper to use, but we're considering these three colours. Which do you like best?
and so on.

Performance Testing

This way of pretesting can be useful for a practical handbook or instructions for activities or skills. Performance testing means that you watch the reader using the skills described in the manual, either in his normal work or as a special test. If he performs the task correctly then the manual has communicated satisfactorily; if not, the reasons why must be investigated, and the manual revised.

Chapter 7 Forming the Message

Let us look at an example of simple pretesting. Picture (a) was in a booklet produced during a workshop in Orissa, India. It was designed for shopkeepers and stall-holders selling food, aiming to persuade them to wash their hands properly and at the right times. This was the first draft of the picture, and was intended to show the owner of a food shop washing his hands before selling sweets to a child and his mother.

(a)

(b)

Picture (a) was pretested in a field survey. Although two-thirds of the respondents identified the shop, there were difficulties in understanding that the person washing his hands was the seller, and the woman with her child a prospective buyer. Obviously the message — that shopkeepers who handle food should have clean hands — was being missed largely because key parts of the picture were not being identified as the artist had intended.
In their discussion of the pretest results, the production team realized the need to cut out distracting background details and to associate the shopkeeper much more directly with his shop by having him seated in front of his wares. Additional clues were planned to help the viewer identify the shop — a sign board and a pair of scales. The figures of the woman and child should not, it was decided, be partially cut off by the picture's edge. The child should be pointing at the sweets and the mother should carry a shopping bag to give a clearer idea of their intentions.

The revised version — picture (b) — incorporated all the suggested changes. It was tried out on a number of shopkeepers and was found to be clearly understood. One problem remained, however: nobody was able to identify the soap in the picture. The production group decided to turn this difficulty to advantage, and to use it as a means of placing extra emphasis on the need to use soap when washing hands, and so they added the caption at the bottom of the page: 'Is the shopkeeper using soap?'

Information from Pretesting

Pretesting can give information on most of the qualities that help material to communicate well.

- **Attractiveness**

Is it appealing and interesting enough to attract and hold people's attention? Are people interested in the subject matter? Does the material tell them things they want to know?

- **Clarity (and understanding)**

Can people read the words and see the pictures clearly? Can they recognize and understand the words and pictures? Is the message clearly understood?

- **Relevance**

Do people feel it relates to them personally, or is it only 'for other people'?

- **Acceptability**

Is it acceptable, or does it contain anything people feel is untrue or offensive?

- **Persuasiveness**

Can it persuade people to do what the communicator requires? Pretests cannot usually show this; it is more often tested later on by evaluation.

Some of this information will overlap with facts discovered during target audience surveys (see page 17). Certainly, the more you have already been able to find out about your target audience, the easier your pretesting should be.

Methods

There is no standard formula for pretesting, since every pretest grows out of the particular circumstances and constraints involved. Methods can range from simple to complex. Simple ones can give useful and valid information, and are often all that is needed. Even the most straightforward pretest with just a few respondents is better than nothing at all.

Let us look at what is probably the most useful and straightforward method of pretesting — individual interviews, using a questionnaire. This method is economical and needs no special facilities. It can be used with any audience, including those who cannot read or write. It can be carried out by people without extensive training in testing and research. It involves the following tasks:

1 Produce a trial or draft version of the material (or part of it). This should not be so well finished and 'polished' that it is very costly and nobody wants to revise it. On the other hand, if it is too rough it may not be fairly tested. (For more information on producing draft material see Book 2.)

2 Identify a suitable group of respondents — a sample group from the intended target audience. If possible, pretest in two different places — this will give a more accurate picture of people's reactions. It is best to limit your target audience to one particular kind of person. If it has to be diverse, try to represent all the main sub-groups in your chosen sample. The size of your sample group will probably depend on the resources available — the larger it is, the more confidence you can have in the results obtained. A sensible number of respondents would be 25-50, but try to interview 15-25 from each main sub-group if you have a diverse target audience.

Chapter 7 Forming the Message

3 Design, check and duplicate a questionnaire. It is worth trying it out on a few people before duplication. Check that the questions can be clearly understood and will give you the facts you need.

4 Select interviewers and brief them. They should be of similar (or slightly higher) status to the people they will be interviewing, and should dress comfortably and simply so that they do not appear too formal or intimidating. Show them the material to be pretested and tell them about the background of the project, the reasons for pretesting and so on. Get them to hold a practice interview and try out the questionnaire — see that they are friendly yet 'neutral', and that they can fill in the questionnaire accurately.

5 Arrange a field test so that the sample group can use the material. This will need time and money. Estimate that a single interviewer can complete about ten short interviews in an eight-hour working day; the results of these interviews may then take another day to tabulate. Costs will include any fees or expenses, transport where necessary, the cost of producing the questionnaire and providing clipboards and pencils.

6 Collect feedback from the sample group by conducting interviews and recording responses.

7 Tabulate these results and draw from them the information you need.

8 Use this information to help revise the material and make it more effective.

The tasks listed above are similar to those undertaken for a target audience survey — look back at the information on pages . . . to . . . Arranging a field test will need the same patience and sensitivity; the sample group must feel confident that they can voice their real opinions without giving offence. Interviews and questionnaires must be planned with the same care.

Using a Questionnaire

Remember to avoid asking leading questions or showing people in any other way how you want them to answer. Do not rely on a respondent's own report of whether he has understood something — he will probably tell you yes, he has, and thus encourage you to feel over confident about the material.

It is a good idea to test different elements before testing the piece of material as a whole. For example, you might show some of the pictures from a booklet on separate pieces of paper first. A basic sequence of questions might start by looking at the illustrations, go on to words or text and finish with more general information, as follows:

Pictures

1 *What do you see in this picture?*

(a standard opening question, which is neutral and encourages the interviewee to explore the picture and describe it in his own words)

2 *What is this?*

(repeat this question, pointing to different details in the picture; a back-up question to test the answers given to the first question and check people's understanding)

3 (If people are recognized in the picture) *Do these people remind you of your friends, or are they different from your friends?*

3a If they are different, *in what way?*

4 *Can you find anything in this picture that is not clear?*

4a If yes, *what?*

5 *How could we make it better?*

6 *Is there anything that we forgot to put in the picture?*

6a If yes, *what?*

Depending on the message and the material you could ask further questions about style, colour, underlying meaning or implications of the subject shown.

Words

Respondents can be asked to read through selected sentences underlined in colour. You could ask them to read a passage out loud in order to see how easily they can read the words. You will need to find out whether they have understood the meaning by asking questions such as:

7 *Please tell me, in your own words, what is the idea written here?*

Ask questions about the size and style of lettering, vocabulary, layout and so on, and watch to see if the reader finds anything difficult to follow when he is reading.

General

When people have had a chance to look through the whole booklet, ask:

8 *Do you think this booklet is asking people to do anything in particular?*

8a If yes, *what?*

9 *Can you find anything in the booklet which you don't think is true?*

9a If yes, *what?*

10 *Can you find anything in the booklet which bothers you or which might offend your friends and neighbours?*

10a If yes, *what?*

11 *Is there anything you particularly like about the booklet?*

11a If yes, *what?*

12 *Is there anything you particularly dislike about it?*

12a If yes, *what?*

13 *In your opinion, how could we make this booklet better?*

Some of the questions are 'closed' questions — designed so that there is a limited number of possible answers, for example: yes/no/don't know, or: very good/good/poor. The possible answers should be written on the questionnaire, so that everything is clear. The response given can quickly be marked or circled by the interviewer:

> **11** *Is there anything you particularly like about the booklet?*
>
>
> (YES)/NO/DON'T KNOW

Alternatively, a code number can be given to each response listed, so that this number can be quickly written down:

Question	Answer recorded as a code number
12 *Is there anything you particularly dislike about it?* 1 = YES 2 = NO 3 = DON'T KNOW	2

'Open' questions could have any number of possible answers. The interviewer cannot know what these answers will be, and has to write down exactly what respondents say, for example:

In your opinion, how could we make this booklet better?

Possible responses:

'Use more stories'
'Make the writing larger'
'Include some more questions and answers'
'I prefer coloured pictures'.

It is a good idea to repeat the question or put it a different way before accepting a 'don't know' answer. People sometimes say 'I don't know' to avoid puzzling over a question or saying what they really think, and some encouragement might help them to respond, for example:

Is there anything that we forgot to put in the picture?

'I don't know'

Maybe we forgot to show some detail that would make the picture easier for people to understand. Please would you look over it again?

'Oh — the woman doesn't look right.'

Why doesn't she look right?

'She should be wearing jewellery.'
Having established a 'yes' or 'no' answer, you may need to find out more, again by probing for further information:

Is there anything that we forgot to put in the picture?

'Yes.'

Chapter 7 Forming the Message

What things did we leave out?

'The mother in the picture should be wearing bracelets.'

You will also need to collect some information about the person being interviewed, for example their age, sex, marital status, number of children (if any), years of schooling and current occupation. Again, look back at the information on target audience surveys (pages 17 to 29).

Analysing Findings

It saves time to use codes or numbers to identify things. Number each completed questionnaire form — they will be easier to sort through if you need to refer back to one later. The responses in the questionnaire can each be given a code; this is essential for open questions, and we shall consider them in a moment. First let us look at closed questions. Responses can be counted quite simply from the markings on the questionnaires, or from the code numbers written down. You will be able to write down totals such as:

Response	Code	Number counted
YES	1	19
NO	2	3
DON'T KNOW	3	8

It is more complicated to count responses to open questions. The first task is to group similar responses together, to reduce the number of different ideas you are dealing with. Taking question 11a, let us arrange the responses into groups. First list all the responses:

11a *What do you like about the booklet?*

Responses listed:

a) the family in the story
b) the pictures
c) the story was interesting
d) it was all good
e) I could read the words nicely
f) the parents were wise
g) I learnt a lot
h) I don't remember exactly

Then decide what categories will suit these different responses, and give a code number to each category:

Categories	**Code Number**
— the story	1
— the characters	2
— the pictures	3
— the lettering	4
— don't know or non-specific	5

You will be able to write down totals such as:

Response (code)	Number counted
1	7
2	4
3	5
4	1
5	11

Go through all the questionnaires and make sure that:

— responses to closed questions are marked, or a code number noted down
— responses to open questions have all been marked with the code number given to that kind of response.

The next task is to make up a tabulation sheet where you can write down all the totals when you count the responses.

Tabulation Sheet				
No.	Questions, responses and codes	Counting responses	Number counted	Percentage of the total response
11	*Is there anything you particularly like about the booklet?* 1 = yes 2 = no 3 = don't know	ℍℍℍℍ ℍℍ ℍℍℍ	19 3 8	63 10 27
11a	*What?* 1. story 2. characters 3. pictures 4. lettering 5. don't know 6. not applicable (answered No. 11 by response 2 or 3)	ℍℍ ℍℍ ℍℍ ℍ ℍℍ ℍℍℍ	7 4 5 1 2 11	23 13 16 3 7 37

Make sure that you have not forgotten to record any of the responses. Check that the responses to each question add up to the right total — in this example the total is 30 interviews.

The results may be clear — if the majority (70 per cent or over) of the respondents like the booklet, it is probably attractive enough as it is; if few people like it (less than 40 per cent), it needs revising. In our example, 63 per cent of the respondents liked the booklet. When results are in this middle range, the pretester may want to look more closely at which respondents hold which opinions. This is done by cross tabulation, using variables such as male/female, literate/non-literate, single/married and so on. Here we shall investigate how the age of respondents might affect their opinions. Draw a grid like the one shown, in which to record the information. Then go through the pile of

Chapter 7 Forming the Message

questionnaires, noting the identification number of each in the appropriate box on the grid.

Did respondents like the booklet?

	Up to age 25	26 and over
Yes	03,06,09,12 15,17,20,21 23,28,30	01,02,05,08 14,22 27,29
No or Don't Know	24	04,07,10,11 13,16,18,19 25,26

⇨

11	8
1	10

These results seem to suggest that people under 25 liked the booklet reasonably well and that older people did not like it so well.

Other parts of the questionnaire might yield more information, or the pretester/producer might wish to investigate further. A larger sample group would, of course, give more clear-cut results.

A further point to consider is how these results relate to the different qualities being tested — the attractiveness, clarity, relevance and acceptability of the material, as mentioned earlier. The following list shows how these qualities are covered by the questions given on page 83 and 84.

Qualities etc.	Questions
attractiveness	11, 11a
clarity (and understanding)	1, 2, 4, 4a, 6, 6a, 7, 8, 8a
relevance	3, 3a
acceptability	9, 9a, 10, 10a, 12, 12a
other points (including recommendations)	5, 13

The totals and percentages for associated questions can therefore be linked together to give a summary of results. When reporting the final results of a pretest, you will probably need to work further on this. By dividing the closed and open questions you can present a group of percentage replies to key questions, and a supporting group of more detailed responses, for example:

Qualities etc.	Key questions	More detailed responses
attractiveness	11	11a
clarity (and understanding)	1, 4, 6, 7, 8	2, 4a, 6a, 8a
relevance	3,	3a
acceptability	9, 10, 12	9a, 10a, 12a
other points		5, 13

Three basic tables might be the best way of presenting results briefly and clearly in a final report:

— the percentage replies to key questions

— more detailed responses which explain the reasons for the percentage replies

— information about the respondents interviewed.

Base your written summary of findings and conclusions on the information given in these tables. Remember that the purpose of pretesting is to help the producer revise material so that it communicates more effectively. Results must be objective, clearly set out and well understood. They must also be used — a report put to one side and left on a shelf will not help anybody.

It may be easier to build a final evaluation into production of materials than to arrange pretesting, but pretesting will usually be more valuable. It is rather late to find that material is not communicating effectively *after* it has been produced and distributed; it will certainly be difficult and expensive to put things right.

Some Final Comments

This book has introduced some of the thinking, planning and preparation that should take place before any material is produced. Hopefully you will now feel able to collect useful information about target audiences, to plan messages with your audience in mind and to arrange a simple pretest. If you would like more information about any of these areas, consult the bibliography. The Trainers' Handbook has more details about books which are recommended, and some addresses which may be useful.

If you are ready to read on:

Book 2: Designing and Producing Artwork

— deals with the design and production of the completed original artwork which is then ready to be printed; it covers information needed both for decision-making and for the practical work involved;

Book 3: Printing Processes

— discusses a range of processes likely to be appropriate to readers; it sums up information which will help in choosing an appropriate method;

Book 4: Managing Resources

— shows how planning and organizing can help the production process to run smoothly, and discusses some aspects of sending work to professional printers.

Glossary

Accessible
within reach

to Acquire
to gain or learn

Adequate
enough, sufficient or competent

Aim
a general idea of one's intentions

to Analyse
to study each part of something carefully, and then probably to compare the results and sum them up in a statement or table

Ambiguous
having more than one meaning

Appraisal
an estimation or judgement of something's quality or value

Appropriate
suitable

Assessment
an estimation or judgement of something's quality or value

Assumption
to make an assumption is to take something for granted

Attainable
can be reached or gained by hard work or effort

Audience
those who see or hear a particular message, presentation or materials prepared for them, the person or people who receive a communication message

Authoritative
having the weight or influence of authority

Case Study
an example used to help illustrate points or principles

Category
a class or group

Channel of Communication
the means by which a message is sent

Characteristic
a distinctive quality

Chronological Order
the order in which things happen, or in which they should be done, from beginning to end

Classified
arranged in classes or groups

Colloquial
language as used in common conversation, chatty, probably using local dialect, idiom or slang phrases

to Communicate
to have something in common with another, to transfer an idea or send a message

Communication Network
existing links or lines of communication in a community or group

Component Parts
the elements or parts that make up the whole

Concept
a general notion, an idea or thought

Concrete
something actual, material, practical or real (as opposed to 'abstract' or 'theoretical')

Considerable
a great amount, important

Consistent
reliable, fixed, free from fluctuations or contradictions

Constraint
a condition that restricts or limits

Contaminated
infected or polluted by contact with germs

Controversial
relating to a dispute or argument

to Convey
to communicate or transmit an idea or message

Credibility
how much a person or statement can be trusted, believed or relied on

Decode
to translate a received message into an idea which is clearly understood (see Encode)

to Demonstrate
to teach, explain or show by practical means

Distribution Channels
the ways in which a message or materials are divided and sent out among those who are to receive them

Encode
to translate an idea into a message in some form (e.g. written words) which can be sent to an audience; the message will have to be decoded by the receiving audience (see Decode)

Environment
surroundings, conditions influencing development or growth

to Evaluate
to judge the value of a project or piece of material by its results

Evident
obvious or clearly to be seen

Exhibition
a presentation, display or public show

Feedback
a response to a message by those receiving it, as observed and used by the communicator

Field Trip
a visit to the place where something is happening or being done, to see the real thing in action

Flannelgraph
a display board covered with hairy or rough-textured cloth, to which cut-out words or pictures backed with similar rough cloth or sandpaper will stick

Flash Cards
word or picture cards that are held up quickly to illustrate a point or encourage immediate response from a group

Guidelines
list of points to help people do

things in the best or easiest way

Inaccessible
see Accessible

Inappropriate
see Appropriate

Indigestible
(of information) not easy to understand

to Innovate
to introduce something new, to make changes

Leaflet
a small folded sheet of paper on which information is printed or duplicated, usually cheaply produced and distributed in large numbers

Mass Media
media used to communicate a message to many people at the same time, e.g. radio, TV, newspapers

Medium (pl. Media)
means or agency used to communicate a message

Message
any communication (e.g. information or an idea) sent from one person to another

Motivation
a feeling of conviction and positive wish to act

Non-projected Visual Aid
a visual aid which does not need any kind of projection equipment (film projector, slide projector etc.) e.g. poster, wallchart, picture card, leaflet, model

Objective
a specific statement of intention, e.g. saying exactly what someone should know or be able to do after they receive a given message

Objectivity
having opinions that are uncoloured by one's own personal feelings or emotions

Oral Rehydration Mixture
water with sugar and salt added in the right proportion — given to children with diarrhoea to replace the liquid their bodies have lost

Originator
the person who first starts something

Participation
taking part in an activity, sharing in something

Percentage
a part of something expressed in hundredths, e.g. $\frac{25}{100}$ of 3200 is 25 per cent or 25% of 3200, which is 800

Perspective
the art of drawing objects on a flat surface so that they appear to have depth, and look solid and realistic

Predetermined
decided or settled beforehand

Preliminary
introducing or preparing for something else that comes later

Pretesting
testing material to check that it is appropriate and effective, before producing the final version or distributing multiple copies

Priority
something that has priority comes first — it is more urgent or more important than other things

Projected Visual Aid
a visual aid that uses equipment such as a film projector or slide projector

Questionnaire
a prepared set of questions asked in order to gain specific information

to Reinforce
to strengthen or support.

Resource
something that enables one to act or do, e.g. money, equipment, materials, manpower

Respondent
someone who answers, e.g. answers the questions put to them by an interviewer with a questionnaire

Response
answer, e.g. answer given to an interviewer's question

Silhouette
a shadow-outline filled with black, giving a bold, solid shape

Simulated Experience
an imitation of real life; circumstances are arranged so that learners can take decisions and try things out without having to face the real consequences of those decisions

Source
the person or group of people sending a message

Stimulating
exciting; something that rouses one to action

Survey
a collection of data; a 'target audience survey' is the collection of information about people to increase one's understanding of the intended audience group

Susceptible
susceptible to something means easily affected by it

Tabulate
to present information in the form of a table

Tangible
can be touched

Target Audience
the intended audience or receiving group for a given message

Terminology
technical terms

Three-dimensional
having three dimensions — height, width and depth; a shape which is solid (as

opposed to a shape drawn on paper, which is 'two-dimensional' and flat)

Topic
subject

Topic Research
research into the subject matter of a piece of communication, to check that the information is accurate, clearly presented, acceptable to the audience and appropriate for their needs

to Transmit
to send or pass on, e.g. a message

Two-way Communication
a situation where the communicator sending a message encourages and takes note of the feedback from his audience, so that they in turn are communicating with him; communication has more chance of being effective it if is two-way

Variable
a factor which can change

Visual
concerned with the sense of sight, can be seen

to Visualize
to make visible, to call up a clear picture of something

References

We gratefully acknowledge permission to use the following material, some of which is copyright.

Chapter 2
1 Information from *Visual Symbols Survey: Report on the Recognition of Drawings in Kenya* by Bernard Shaw, CEDO, 1969.
2 Example based on a survey from *Understanding Pictures in Papua New Guinea* by Bruce L. Cook, the David C. Cook Foundation, 1980.

Chapter 7
1 Pictures from a Kenyan survey quoted in *Applied Communication in Developing Countries* by Andreas Fuglesang, Dag Hammarskjold Foundation, 1973.
2 Picture from *Visual Symbols Survey: Report on the Recognition of Drawings in Kenya,* Bernard Shaw, CEDO, 1969.
3 Pictures from surveys quoted in *Understanding Pictures* by David Addison Walker, University of Massachussetts, 1979.
4 Pictures from a book about Nazar (diarrhoea in children) for village health workers; produced by Maeve Moynihan and others at the department of Preventive and Social Medicine, Banares Hindu University, 1980.
5 Example based on information given in *The Essentials of First Aid,* a St. John Ambulance booklet.
6 Examples 2-4 based on information given in *A Guide to Growing Fruit and Vegetables,* an Extension Aids Publication, Malawi, 1973.
7 Diagrams from *Appropriate Technology*, Vol. 8 No. 4, 1982.

Bibliography

Note: Publishers addresses are given in the Trainers' Handbook in this package; so, also, are further details about some of these books that may be especially useful.

Abbatt, F. R. *Teaching for Better Learning: A Guide for Teachers of Primary Health Care Staff,* WHO, 1980, (137 pp.)

Bertrand, Jane T., *Communications Pretesting,* The Community and Family Study Centre, The University of Chicago, USA, 1978 (144 pp.) ISBN 0 89836 006 4

Cook, Bruce L., *Understanding Pictures in Papua New Guinea,* David C. Cook Foundation, USA, 1980 (113 pp.)

Coombs, Philip H. and **Ahmed, Manzoor,** *Attacking Rural Poverty: How Nonformal Education Can Help,* A World Bank Publication, 1974 (292 pp.) ISBN 0 8018 1601 7

Coppen, Helen, *Wallsheets: Choosing, Using, Making,* National Committee for Audio-Visual Aids in Education, UK, 1971

Crone, Catherine D. and **St. John Hunter, Carman (Eds.),** *From the Field: Tested Participatory Activities for Trainers,* World Education, USA, 1980 (148 pp.) ISBN 0 914262 19 X

Feliciano, Gloria D., *Research in Population Communication,* UNESCO, 1978 (85 pp.) ISBN 92 3 101512 5

Fuglesang, Andreas, *Applied Communication in Developing Countries,* Dag Hammarskjold Foundation, Sweden, 1973 (124 pp.)

Fuglesang, Andreas, *About Understanding: Ideas and Observations on Cross-Cultural Communication,* Dag Hammarskjold Foundation, Sweden, 1982 (231 pp.) ISBN 91 85214 09 4

Fussell, Diana and Haaland, Ane, *Communicating with Pictures in Nepal,* UNICEF, 1975 (22 pp.)

Goldsmith, Evelyn, *Research into Illustration,* Cambridge University Press, UK, 1984 (487 pp.) ISBN 0 521 25674 7

Holmes, Alan C., *A Study of Understanding of Visual Symbols in Kenya,* OVAC, UK, 1963 (32 pp.)

Holmes, Alan C., *Visual Aids in Nutrition Education,* FAO, 1968 (154 pp.)

An International Family Planning Project publication: *Working with Villagers,* The American Home Economics Association, International Family Planning Project, USA, 1981 (3 handbooks)

Jarmul, David, *Plain Talk: Clear Communication for International Development,* VITA, USA, 1981 (75 pp.) ISBN 0 86619 131 3

Jenkins, Janet, *Materials for Learning,* Routledge and Kegan Paul Ltd., UK, 1981 (208 pp.) ISBN 0 7100 0808 2

Jenkins, Janet, *Mass Media for Health Education: IEC Broadsheets on Distance Learning: 18,* International Extension College, UK, 1983 (67 pp.)

Leonard, Ann, *Sin Palabras: Printed Materials for People Who Do Not Read,* A Cycle Publication, USA, 1979 (12 pp.)

A Lesotho Distance Teaching Centre Publication:
Understanding Print, Distance Teaching Centre, Lesotho, 1976 (60 pp.)

McBean, George (Ed.), *Illustrations for Development,* Afrolit Society, Kenya, 1980 (69 pp.)

Mitton, Roger, *Practical Research in Distance Teaching: a Handbook for Developing Countries,* International Extension College, UK, 1982 (321 pp.) ISBN 0 903632 24 1

Parlato, Ronald; Parlato, Margaret Burns and **Cain, Bonnie J.,** *Fotonovelas and Comic Books: the Use of Popular Graphic Media in Development,* Office of Education and Human Resources, Development Support Bureau, AID, USA, 1980 (243 pp.)

Pett, Dennis W. (Ed.), *Audiovisual Communication Handbook,* World Neighbors, USA, (125 pp.)

Rawson-Jones, Daphne and **Salkeld, Geoffrey (Eds.),** *Communicating Family Planning,* IPPF, UK, 1972 (195 pp.)

Roppa, Guy M. and **Rodriguez de Andrade, Rosa V.,** *Community Participation in Family Health,* Ministry of Public Health, Ecuador, and ODA; UK, 1980 (187 pp.)

Saunders, Denys J., *Visual Communication Handbook,* Lutterworth Press, UK, 1976, 2nd impression (127 pp.) ISBN 0 7188 2083 5

Scandlen, Guy B; Herm, Jim and **Pape, William R.,** *Communication Strategy Planning Workbook,* UNICEF, Thailand, 1979 (folder of training materials)

Scotney, N., *Health Education,* AMREF, Kenya, 1976 (141 pp.)

Shaw, Bernard, *Visual Symbols Survey: Report on the Recognition of Drawings in Kenya,* CEDO, UK, 1969 (54 pp.)

Varma, R; Ghosal, S. L.; Bowers, J. and **Hulls, R. H. (Eds.),** *Action Research and the Production of Communication Media,* Agricultural Extension and Rural Development Centre, University of Reading, UK, 1973 (32 pp.)

Walker, David A., *Understanding Pictures,* Centre for International Education; University of Massachusetts, USA, 1979 (380 pp.) ISBN 0 932288 55 3

Werner, David and **Bower, Bill,** *Helping Health Workers Learn,* The Hesperian Foundation, USA, 1982 (632 pp.)

Zimmer, Anne and **Zimmer, Fred,** *Visual Literacy in Communication: Designing for Development,* Hulton Educational Publications Ltd., UK, and the International Institute for Adult Literacy Methods, Iran, 1978 (144 pp.) ISBN 0 7175 0806 4

www.ingramcontent.com/pod-product-compliance
Ingram Content Group UK Ltd.
Pitfield, Milton Keynes, MK11 3LW, UK
UKHW051822210426
5322IPUK00023B/597